Engaging Indonesia

Critical Dialogues on Culture and Society

Engaging Indonesia brings together current debates in social and cultural studies relating to the study of Indonesia, with the aim of giving voice to both leading and emerging Indonesianists. Critical and interdisciplinary, the books in the series contribute to knowledge formation by addressing pressing issues facing the country—cultural, social, and political—from contemporary and historical perspectives. Works may include research-based monographs, edited volumes, handbooks, and translations of both recent, and seminal/historical works. The series is particularly welcoming of topics relating to issues of urban and environmental change, to questions of identity and place, to problems of governance and conflicts, to the politics of representation, language, and performances, and to the global and postcolonial contexts in which these are addressed. Special consideration is given to studies that cover underrepresented regions and marginalized voices, as well as those that seek to recover lost archives, uncounted political events, or forgotten historical figures. Relevant to students and scholars from a broad range of disciplines in the humanities and social sciences with a particular interest in researching Indonesia, the series champions intellectual engagement with the archipelago's rapidly changing landscape—from within, and beyond.

Abdul Halim Iskandar

Accelerating the Achievement of the National Sustainable Development Goals in Indonesia

SDGs Desa (Volume 1)

Springer

Abdul Halim Iskandar
Jakarta, Indonesia

ISSN 2948-2909 ISSN 2948-2917 (electronic)
Engaging Indonesia
ISBN 978-981-97-8918-4 ISBN 978-981-97-8919-1 (eBook)
https://doi.org/10.1007/978-981-97-8919-1

Jointly published with Yayasan Pustaka Obor Indonesia
The print edition is not for sale in Indonesia. Customers from Indonesia please order the print book from: Yayasan Pustaka Obor Indonesia.

Translation from the Indonesian language edition: "SDGs DESA: Accelerating The Achievement of National Sustainable Development Goals (Versi Bahasa Inggris)" by Abdul Halim Iskandar, © Pustaka Obor Indonesia 2022. Published by Pustaka Obor Indonesia. All Rights Reserved.

Iskandar, A. Halim. SDGs DESA: Accelerating the Achievement of National Sustainable Development Goals (Versi Bahasa Inggris). Jakarta: Yayasan Pustaka Obor Indonesia, 2022. ISBN 978-623-321-145-1.

This Springer imprint is published by the registered company Springer Nature Singapore Pte Ltd.
The registered company address is: 152 Beach Road, #21-01/04 Gateway East, Singapore 189721, Singapore

Preface

When I was named Minister of Villages, Development of Disadvantaged Regions, and Transmigration by President Joko Widodo, on October 23, 2019, I felt it was an extraordinary mandate and one that would entail significant challenges. I would have to solve development problems in 75,265 villages, 270 rural areas, 62 disadvantaged areas, and 619 transmigration areas throughout Indonesia, quickly, productively, and oriented toward tangible results. That was part of the President's message when he introduced us, members of the Advance Indonesia Cabinet, to the Indonesian people.

In fact, life in the village, with all its dynamics, is very much a part of me. I was born and grew up in the village of Denanyar in Jombang regency, East Java province. After graduating from the Yogyakarta State Institute of Teaching and Education (now Yogyakarta State University) in 1987, armed with knowledge and experience gained from Islamic boarding school and my years on campus, I decided to live back in the village of my birth, so I could explore the lives of villagers, and together seek solutions to their problems.

In 1987, when I returned to my village, the number of poor people in Indonesia's rural village areas was more than 20 million, while about 9 million were living in poverty in urban areas. By March 2020, according to data from Statistics Indonesia (BPS), the number of poor people in Indonesia amounted to 26.42 million or 9.78 percent of the total population. In rural village areas, there were 15.26 million poor people—a poverty rate of 12.82 percent. In urban areas, there were 11.16 million poor people—a poverty rate of 7.38 percent.

Considering this data, I have concluded the root of development problems in Indonesia lies in the village. The village is a source of problem identification; the highest poverty level is in the villages, most people with low health rates are in the villages, low purchasing power and low education levels are still synonymous with villagers. For this reason, I believe that in order to improve the quality of life of the Indonesian people, to realize an Advanced Indonesia, one must start from the village.

Since the implementation of Law No. 6 of 2014 on Villages, the government has channeled village funds as a stimulus for villages to solve their development problems by themselves. In 2015, the government disbursed Rp20.67 trillion in village funds. The amount was increased to Rp46.98 trillion in 2016, then to Rp60 trillion in both

2017 and 2018, to Rp70 trillion in 2019, to Rp71.19 trillion in 2020, to Rp72 trillion in 2021, while the planned fund for 2022 reached Rp68 trillion.

So far, it must be admitted, the use of village funds has not been optimal in solving the problems that exist in the village. There is no single reason for this; factors range from village development planning tools to the quality of village human resources. For this reason, a framework for the direction of village development policy was needed, with a focus oriented toward tangible results.

In view of these challenges, I initiated the concept of SDGs Desa (Village Sustainable Development Goals) as an integrated effort to accelerate Indonesia's achievement of the United Nations' 17 Sustainable Development Goals (SDGs), as regulated in Presidential Regulation No. 59 of 2017. Who would not be interested in the most complete development concept to date? The global SDGs include all development conventions practiced in the world, such as economic, social, environmental and cultural development, disaster mitigation, and so on. The 17 SDGs also aim to ensure that no one is left behind, meaning that all people must be able to benefit from the results of development.

In establishing the SDGs Desa, I added one sustainable development goal that is unique to Indonesian villages. So the 18 SDGs Desa do not just localize the global SDGs, but place them as an integral part of daily village life in Indonesia. The 18th SDGs Desa is: Dynamic Village Institutions and Adaptive Village Culture. This additional goal is built around the local wisdom of villages in Indonesia, so that development is truly and deeply interpreted and conducted as an improvement in village life.

With these 18 SDGs Desa, village development, whether carried out by the village government, or through interventions by the supra-village (regency, provincial or central) governments, will be more focused and have a clear direction in accordance with the real conditions of the village, and include all villagers, with no one left behind.

A planned trilogy of books on the SDGs Desa will comprise:

1. A book on the concepts and ideas of the SDGs Desa, namely this book.
2. A book on the methods of measuring the SDGs Desa.
3. A book on the results of the SDGs Desa.

This book is in the hands of readers because of the blessing of prayers from the author's mother, Mrs. Muhassonah Hasbullah (Nyai H. Iskandar), the sincerity of his beloved wife, Lilik Umi Nasriyah, along with our children, in-laws, and grandchildren, all of whom have sincerely spent a lot of time with the author to assist in the creative and critical processes of putting his thoughts and ideas into a book worth reading. Hopefully this endeavor will be an inspiration to others, and become a 'good deed' that continues to reap rewards until *Yaumil Qiyamah* (Judgement Day).

To the many people who have helped with the publication of this book, the author is extremely grateful. May God reward you with much kindness.

As a son of the village, whose work is reflected in his dedication, ideals, solemnity and struggles, the author sincerely dedicates this book to all villages throughout

Indonesia. The author realizes this book may contain some shortcomings. Nevertheless, it is hoped this book will serve as a dynamic and constructive discussion material for a building a better future for Indonesia's villages. Therefore, the author expects feedback and suggestions from readers for improving future editions of this book.

It is the author's sincere hope that the English edition of this book will provide foreign governments, development organizations and partners with greater insight into how Indonesia is adapting the SDGs to best meet the needs of the nation's villagers.

Happy reading.
Wallahul Muwaffiq Ilaa Aqwamith Tariiq
[Allah is the one who guides to the straight path].

Jakarta, Indonesia
February 2022

Abdul Halim Iskandar

Contents

About the Author

Abdul Halim Iskandar has served as Minister of Villages, Development of Disadvantaged Regions, and Transmigration (2019–2024). He grew up with his extended family at Mambaul Ma'arif Islamic Boarding School in Denanyar village, Jombang regency, East Java province. This Islamic boarding school (*pesantren*) was founded by his great-grandfather, Bisri Syansuri, who was also one of the founders of Indonesia's largest mass Muslim organization, Nahdlatul Ulama (NU). After elementary and high school education at the *pesantren*, he completed his undergraduate studies in philosophy and sociology of education at Yogyakarta State Institute of Teaching and Education (now the State University of Yogyakarta) in 1987. He later received his master's degree in Field Education Management from Malang Teacher Training College (now Malang State University) in 1995.

He is no stranger to village life, having been born and raised in a village. From the time he earned his bachelor's degree until receiving his mandate as chairman of the East Java Provincial Legislative Assembly, most of his life has been devoted to solemnly serving the NU, pesantren, and villages. Having grown up as a village boy, the young Abdul Halim firmly believed that change must be driven by young people, and that Indonesia's development must start from the village. Therefore, as chairman of Jombang Regency Legislative Assembly for two terms (1999–2009), deputy chairman (2009–2014) and then chairman of the East Java Legislative Assembly (2014–2019), he has championed village development, education, health, pesantren, and micro, small and medium enterprises. A policy in the form of a regional regulation on empowering village communities, in both Jombang and East Java, was introduced when he chaired the local legislative assemblies, long before the 2014 Village Law was born. For this reason, he was awarded an Honorary Doctorate in Community Empowerment Management by Yogyakarta State University in 2020, and Honorary Professor in Sociolinguistics by Surabaya State University in 2023.

His shrewdness in political leadership cannot be separated from his roots in the village pesantren and from his time in Yogyakarta, where he was active in the Indonesian Islamic Students Movement (PMII). At NU, he has served as a manager and is now *A'wan* (senior policymaker) of the NU East Java Board (2018–2023). In the National Awakening Party (PKB), he has served for over 10 years on the Regional Board of Directors of the party's East Java chapter (2011–2022) and is one of the chairs of the PKB Central Executive Board (2019–2024).

In educational institutions, his work began when he returned to his home village after completing his studies in Yogyakarta. From 1987 until now, he has served on the board of directors of Mambaul Ma'arif Islamic Boarding School. From 1987–1989 he taught at Mambaul Ma'arif State Aliyah Madrasah (an Islamic high school). He was a lecturer at Hasyim Asy'ary Islamic Institute (Ikaha, now Hasyim Asy'ari University) in Tebuireng, Jombang, from 1987–2004. He served as Dean of the Tarbiyah Faculty from 1993–1997. He initiated the establishment of Sultan Agung Vocational High School in Jombang, where he served as Principal from 1996–2000. Currently, he is Chairman of the Advisory Council of Surabaya State University (2018–2022), and General Chairman of the Advisory Council of Yogyakarta State University (2020–2023).

Together with religious leaders and NU activists, he is active in the health sector. He is one of the founders of the Jombang NU Hospital, where he has also served a Commissioner since 2019. In sharing his thoughts and ideas, he has authored books, as well as numerous articles in the print and electronic media. In 2012, with Thoriqul Haq (now the Regent of Lumajang), he co-wrote Resolusi Jihad Baru: Ijtihad NU Melahirkan PKB (Resolution of the New Jihad:

Ijtihad NU Gives Birth to PKB). An anthology of his media columns was published in 2019 with the title, Negeri Holopis Kuntul Baris (The Land of Working Together to Resolve Big Things). His articles have been published in Kompas, The Jakarta Post, Kontan, Padang Ekspres, Karawang bekasi Ekspres, Jawa Pos, Kedaulatan Rakyat, Republika, and other daily newspapers.

List of Figures

List of Tables

Chapter 1
Village Development Agenda

Abstract Development can be viewed as a post-colonial project that aims to achieve progress in a region.

If you want to make a change,
do not submit to reality,
as long as you are sure you are on the right path.
- Gus Dur -

Development can be viewed as a post-colonial project that aims to achieve progress in a region. Simply put, development is defined as an effort to change from underdeveloped to developed, an effort to improve economic, political, cultural, and community infrastructure (Fakih, 2002). One of the key indicators of development is income per capita.

In Indonesia and many other countries, the general concept of development in the 1960s and 1970s was always seen as an economic phenomenon. According to this development concept, the growth of gross national product (GNP) per capita will result in a trickle-down effect to the wider community in the form of jobs and other economic opportunities (Hikmat, 2001). In reality, the fruits of this capitalist hypothesis did not eventuate. GNP growth was not followed by a trickle-down effect; on the contrary, this development model caused numerous social problems, ranging from inequality and environmental damage to undermining social and cultural aspects of local communities (Tanjung et al. 2018).

The various social problems led to public criticism of traditional development theory and strategy, resulting in the formulation of new development strategies, ranging from human-based development, self-sufficient development, development based on fulfillment of basic needs, as well as sustainable development strategies.

A. H. Iskandar, *Accelerating the Achievement of the National Sustainable Development Goals in Indonesia*, Engaging Indonesia, https://doi.org/10.1007/978-981-97-8919-1_1

1.1 International Community Response

Global discourse on development strategy is highly dynamic because the choice of development model is directly related to people's lives and their right to earn a decent living. Therefore, the problem of how to achieve prosperity and economic sufficiency for all people is a concern of all countries in the international community.

In September 2000, the United Nations Millennium Summit, which was attended by 189 member countries, mostly represented by heads of state and heads of government, made a landmark decision in the history of international development by adopting an inclusive approach for the fulfillment of basic human rights. The Millennium Summit agreed upon and declared the Millennium Development Goals (MDGs) (Widjojo et al. 2004).

The MDGs were developed from the International Development Goals, which were the result of a review of development policies issued by the Development Assistance Committee of the Organisation for Economic Cooperation and Development (OECD-DAC) in the mid-1990s. In addition to revising the International Development Goals into the MDGs, the Millennium Summit also produced several consensus efforts to achieve the MDGs, whose main concerns were human rights, good governance, democratization, conflict prevention, and peace building (Harcourt 2004).

As a collective agreement, the MDGs were the responsibility of all countries that participated in the Millennium Summit, both to their people and jointly between governments. The MDGs had diverse goals and targets, ranging from reducing extreme poverty and hunger to achieving basic education levels, promoting gender equality, reducing child mortality, improving maternal health, combating HIV/AIDS and various other diseases, as well as ensuring environmental sustainability, and forming a global partnership for development.

To facilitate the implementation of programs to achieve the MDGs, supporting programs were conducted with actions that considered and accommodated local values according to the characteristics of each country.

The MDGs comprised 8 goals, 18 targets, and 48 indicators to be achieved by 2015. Each goal had one or more targets accompanied by numerous indicators, compiled by expert consensus from the United Nations Secretariat, the International Monetary Fund (IMF), the Organisation for Economic Development and Cooperation (OECD), and the World Bank (UNDP 2008).

Prior to the declaration of the MDGs at the Millennium Summit in 2000, several international development goals had been set from the 1960s to the 1980s. Some of these development goals were reinforced at global conferences organized by the United Nations in the 1990s, including the 1990 World Summit for Children, held in New York, the United States; the 1990 World Conference on Education for All, held in Jomtien, Thailand; the 1992 United Nations Conference on Environment and Development, held in Rio de Janeiro, Brazil; and the 1995 World Summit on Social Development held in Copenhagen, Denmark (Widjojo et al. 2004). However, the MDGs did not conflict with previous global commitments, because some of the

MDGs had actually been proclaimed in the International Development Goals by developed countries in the OECD in 1996 and were later adopted by the United Nations, the World Bank, and the IMF (UN 2000).

Despite being a signatory to the MDGs at the declaration of the Millennium Summit in 2000, Indonesia did not formally incorporate them into government planning policies until 2010. Nevertheless, Indonesia was committed to achieving the MDGs by 2015. Substantially, the objectives of the MDGs had already been reflected in the now-defunct Broad Guidelines of State Policy (GBHN) and the subsequent National Development Program (Propenas).[1]

Formally, Indonesia's commitment to implement the MDGs was outlined in the formulation of poverty alleviation strategies, at the national and regional levels, which directly related to the achievement of the MDG targets. The legal framework used as the basis for accelerating the achievement of the MDGs in Indonesia was Presidential Instruction No. 3 of 2010. This was addressed to heads of all relevant government agencies and institutions, as well as regional heads, as a reference for implementing equitable development programs for the sustainability and sharpening of National Development Priorities.

1.2 From MDGs to SDGs

At the 60th General Assembly of the United Nations on September 14–16, 2005, an evaluation of the first five years of the implementation of the MDGs was carried out. The evaluation found that 50 countries failed to achieve at least one of the MDG targets. Meanwhile, 65 countries were deemed at risk of completely failing to achieve at least one MDG by 2040.

The deadline for the implementation of the MDGs in all countries, including Indonesia, was December 2015. By that time, Indonesia had achieved 49 of the 67 indicator targets set. Clearly, this achievement demonstrated an increase in the quality and standard of living for the people of Indonesia.

The 2014 *Report on the Achievement of the MDGsin Indonesia* outlined numerous accomplishments (National Development Planning Agency and Statistics Indonesia, 2015), including the proportion of the population whose income is less than USD 1 (purchasing power parity) per capita per day, decreasing from 20.6 percent in 1990 to 5.9 percent in 2008. Gender inequality in education from elementary to tertiary levels had decreased, including the literacy ratio of females to males aged 15–24 years. There was also a decline in the incidence and mortality rate of malaria. In addition, the incidence, prevalence, proportion, and mortality rate of tuberculosis cases decreased, while the number of those treated and cured increased. Furthermore,

[1] Indonesia's Broad Guidelines on State Policy (GBHN) had been implemented in Five-Year Development Plans (Repelita), but the GBHN and Repelita were dropped after the Indonesian Constitution was amended over 1999–2002. They were replaced by the National Development Program (Propenas), which was later replaced by the National Medium-Term Development Plan (RPJMN) and National Long-Term Development Plan (RPJPN).

there was an increase in the proportion of households with sustainable access to safe drinking water in urban areas.

At the end of 2014, several indicators of the MDGs still required hard work to be achieved. The percentage of the population living below the national poverty line reached 11.25 percent, still far from the MDG target of 7.55 percent. The proportion of the population with a calorie intake below the minimum dietary energy consumption level of 1400 kcal/capita/day reached 17.39 percent, far from the target of 8.5 percent; while those with a consumption level of 2000 kcal/capita/day reached 66.96 percent, far from the target of 35.32 percent. Also, the under-five and infant mortality rates, as well as the maternal mortality rate, had all decreased but had not met their targets. Similarly, the prevalence of HIV/AIDS and the proportion of the population aged 15–24 with comprehensive knowledge of HIV/AIDS had not met target levels. In terms of the environment, the amount of carbon dioxide emissions still exceeded the target of the MDGs.

Of course, we should be grateful for the MDG achievements that Indonesia reached, while continuing to make systematic efforts to continuously improve the quality of life and living standards of the Indonesian people. Globally, the end of the MDGs was the starting point for countries of the world to start formulating a new platform for global development as a follow-up and for sustainability of the achievements of the MDGs.

On August 2, 2015, at the United Nations Headquarters in New York, 193 countries unanimously agreed on a new global development resolution, which was titled *Transforming Our World: The 2030 Agenda for Sustainable Development*. The following month, on September 25–27, 2015, at the same location, representatives of 193 UN member states held the Sustainable Development Summit. This meeting agreed to and ratified the Sustainable Development Goals (SDGs), a global development agenda of 17 goals that are specified in 169 targets, which are inclusive and indivisible, global in scope and universally applicable, so that no one will be left behind, and with an achievement period of up to 2030. As a commitment of the international community, the SDGs are a new milestone in the development of countries, continuing the MDGs, for a better quality of human life (UN 2015).

Although a continuation of the MDGs, the SDGs have several differences compared to the earlier set of goals. The SDGs accommodate development problems more comprehensively, both qualitatively, by accommodating development issues that were not included in the MDGs, and quantitatively, with the objective of completing each of the goals and targets set. Likewise, with the formulation process, the SDGs are more participatory and not exclusively bureaucratic and involve non-governmental stakeholders, such as non-governmental organizations, universities and academics, businesses and the private sector, and other interest groups.

The SDGs are guided by five basic principles that balance economic growth, social progress, and environmental protection. Known as the 5 Ps (UN 2015), these principles are:

1. People

This first principle of global development places humans as the main concern, by seeking to end poverty and hunger in all their forms and dimensions. This principle must also ensure that all people can fulfill their needs with dignity and equality, and live in a healthy environment.

2. Planet

This principle directs the agenda of protecting the planet from all forms of degradation and damage. Protection can be done through sustainable production and consumption, sustainable management of natural resources, and by taking urgent and strategic action on climate change to support the needs of present and future generations.

3. Prosperity

This principle aims to ensure that all human beings can enjoy prosperous and fulfilling lives and that economic, social, and technological progress, as well as education and health, occur in harmony with nature.

4. Peace

This principle seeks to foster peaceful, just, and inclusive societies, free from fear and violence. This is crucial because there will be no sustainable development without peace, and there will be no peace without sustainable development.

5. Partnership

This principle is a strategy for implementing and achieving the sustainable development agenda, through mobilizing and enhancing partnerships and collaboration between various parties and the international community. Based on a spirit of strengthened global solidarity, this principle is particularly focused on ending poverty through the participation of all countries and all stakeholders.

As a new development framework, which has accommodated all of the international changes that occurred after the implementation of the MDGs, which ended in 2015, the SDGs are vitally important as a direction of global development.

The SDGs development concept rests on three main pillars of sustainable development. First, social progress, such as human development, education, and health. Second, economic growth, manifested by the availability of facilities and infrastructure. Third, environmental balance, in the form of the availability of natural resources and good quality of the environment. This means the SDGs are a global initiative that aims to create a better human life in social and economic dimensions, through synergy with the environment.

1.3 From Global to Indonesia

The 2030 global development agenda known as the SDGs is much more than a declaration by the UN member states. Although not legally binding, as a joint commitment, the SDGs must be ratified into the national development planning of any country that is a signatory to them.

Also, the spirit of the SDGs should be incorporated into the development policies of the signatories.

Of course, the implementation of the SDGs must involve monitoring and evaluation activities as policy inputs and structured information for policy stakeholders.

As a global initiative that emphasizes the spirit of inclusiveness and participation, the preparation and implementation of the SDGs at national and regional levels cannot be separated from a holistic approach between economic development, social inclusion, and environmental sustainability, while still prioritizing the characteristics and priorities of each country.

Indonesia's commitment to implementing the SDGs was demonstrated by the presence of the country's then-Vice President, Jusuf Kalla, at the United Nations General Assembly for the signing of the SDGs in 2015: mainstreaming the 17 SDGs and 169 targets in the national development agenda, with an inclusive and participatory approach, prioritizing local wisdom, and the principle of no one left behind.

In the same year, a meeting of Indonesia's 'Working Cabinet' decided to prepare a draft legal framework for the implementation of the SDGs via a Presidential Regulation. In early 2016, preparation of the legal framework commenced in an open and participatory manner, with the involvement of the public, universities, and the Coalition of Civil Society Organizations for the SDGs.

The following year, President Joko Widodo signed Presidential Regulation No. 59 of 2017 on the Implementation of the Sustainable Development Goals. This regulation lists the 17 SDGs, regulates the role of each ministry and state institution, as well as the role and involvement of non-governmental stakeholders, such as civil society groups, academics, philanthropic entities, and the business sector, in achieving the SDGs.

In addition, the Presidential Regulation contains three important mandates to be followed up by the central and regional governments. First, the preparation of a National Roadmap for the SDGs, as a planning document that contains strategic policies for the stages of achieving Indonesia's National Development Goals from 2016 to 2030 in accordance with national development targets. Second, the preparation of a National Action Plan for the SDGs, outlining programs and activities of a five-year work plan for the implementation of various activities, which directly and indirectly support the achievement of the SDGs in accordance with national development targets. Third, the preparation of a Regional Action Plan for the SDGs, as a five-year work plan for the implementation of activities, directly and indirectly supporting the achievement of the SDGs in accordance with regional development targets.

The Sustainable Development Goals contained in Presidential Regulation No. 59 of 2017, as a derivative of the SDGs, are as follows:

(1) No Poverty: There is no poverty of any kind throughout Indonesia.
(2) Zero Hunger: Zero hunger, achieving food security, and sustainable agriculture.

(3) Good Health and Well-being: Encouraging and realizing a healthy and prosperous life for all people of all ages.
(4) Quality Education: Ensuring equal distribution of quality education and increasing learning opportunities for everyone.
(5) Gender Equality: Realizing gender equality and empowering women.
(6) Clean Water and Adequate Sanitation: Ensuring the availability of clean water and sustainable sanitation for everyone.
(7) Affordable and Clean Energy: Ensuring access to affordable, reliable, sustainable, and modern energy sources for everyone.
(8) Decent Work and Economic Growth: Supporting and facilitating sustainable economic development, productive employment, and decent work for everyone.
(9) Industry, Innovation, and Infrastructure: Building quality infrastructure, encouraging innovation, and sustainable industrial improvement.
(10) Reduced Inequalities: Reducing inequality, both between people and between regions in Indonesia.
(11) Sustainable Cities and Communities: Building quality, safe, and sustainable cities and settlements.
(12) Responsible Consumption and Production: Ensuring sustainable consumption and production patterns that are environmentally friendly.
(13) Addressing Climate Change: Acting quickly to deal with climate change and its impacts.
(14) Marine Ecosystems: Conserving and maintaining the sustainability of the sea and marine life resources.
(15) Terrestrial Ecosystems: Protecting, restoring, and enhancing the sustainable use of terrestrial ecosystems, and managing them sustainably.
(16) Peace, Justice, and Strong Institutions: Promoting community peace for sustainable development and providing access to justice for all.
(17) Partnerships for the Goals: Revitalize and revive global partnerships, as well as interregional partnerships for sustainable development.

1.4 SDGs Desa Support the SDGs

In the national context, the SDGs, which are global in scope, were ratified in Indonesia as *Tujuan Pembangunan Berkelanjutan* (literally: Sustainable Development Goals). The policies contained in Presidential Regulation No. 59 of 2017 needed to be mainstreamed in all government policies and programs, and among other stakeholders, starting from the national level to the village level. First, the SDGs have been integrated into the national development agenda. Subsequently, many regions, especially at the provincial level and some at the regency/municipal level, have followed up by integrating the SDGs into the regional development agenda.

The question is, what about the integration of SDGs at the village level? In fact, after the enactment of Law No. 6 of 2014 on Villages, the paradigm of village

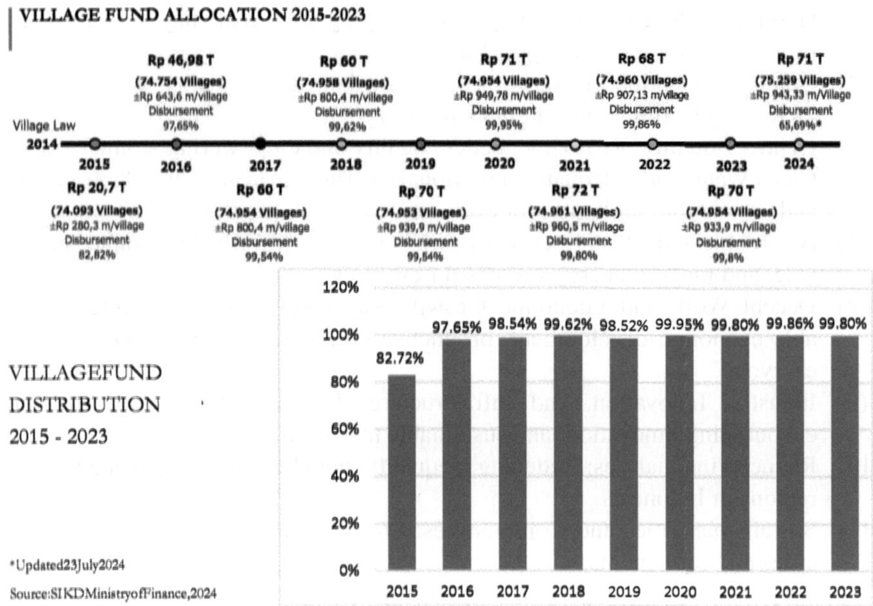

Fig. 1.1 Village fund allocation, 2015–2023. *Source* Ministry of Villages, Development of Disadvantaged Region and Transmigration, 2024

development underwent a drastic change. Where the village unit was previously viewed as merely an object, it is now both a subject and an object of development. One of the indicators of this change can be seen in the increasing allocation of village funds from 2015 to 2019, as shown in Fig. 1.1.

The amount of village funds allocated under Law No. 6 of 2014 on Villages represents vast potential for supporting the achievement of the SDGs. For this reason, villages must be included in the design of mainstreaming sustainable development, as regulated in Presidential Regulation No. 59 of 2017.

In addition to having substantial resources, the potential of villages in supporting the achievement of the SDGs can be seen in two aspects, namely:

1. Territorial Aspect

According to data from the Ministry of Home Affairs (2019), 91 percent of Indonesia's territory is village areas. Given that 11 of the SDGs are closely related to village areas (SDGs 7–17), villages can contribute to 91 percent of the achievement of those 11 SDGs.

2. Population Aspect

Data from Statistics Indonesia (Statistics Indonesia 2020c) shows that 43 percent of Indonesia's population lives in villages. As six of the SDGs are closely related to villagers (SDGs 1–6), the villagers themselves can contribute to 43 percent of the achievement of those six SDGs.

Thus, it can be concluded that mainstreaming village development through an appropriate focus on goals can contribute to 74 percent of the achievement of the National SDGs.

Therefore, mainstreaming village development to support the achievement of the SDGs is of vital importance. This book emphasizes the concept of mainstreaming the SDGs Desa to support the achievement of the SDGs.

References

Fakih, Mansour. 2002. *Runtuhnya Teori Pembangunan dan Globalisasi* (The Collapse of Development and Globalization Theories). Yogyakarta: Insist Press and Pustaka Pelajar.

Harcourt, W. 2004. *The Road to the UN Millennium Development Goals: Some Insights into the International Debate.* Accessed from www.ncdo.nl.

Hikmat, R.H. 2001. *Strategi Pemberdayaan Masyarakat* (Community Empowerment Strategy). Bandung: Humaniora.

National Development Planning Agency (Bappenas) and Statistics Indonesia (BPS). 2015. *Indeks Pembangunan Desa (IPD) 2014: Tantangan Pemenuhan Standar Pelayanan Minimum Desa* (Village Development Index 2014: Challenges in Meeting Minimum Village Service Standards). Jakarta: Bappenas and BPS.

Statistics Indonesia (BPS). 2020c. *Statistik Indonesia Tahun 2020* (Indonesian Statistics 2020). Jakarta: BPS.

Tanjung, M. A., et al. 2018. *Arus Baru Ekonomi Indonesia.* Jakarta: Flasmediatama.

United Nations. 2000. *Progress towards the International Development Goals: A Better World for All.* Washington: UN.

United Nations. 2015. *Transforming Our World: The 2030 Agenda For Sustainable Development.* New York: UN.

UNDP. 2008. *Target and Indicators.* New York: UNDP.

Widjojo, Prasetijono et al. 2004. Indonesia. *Laporan Perkembangan Pencapaian Tujuan Pembangunan Milenium (Millennium Development Goals).* Jakarta. Bappenas.

Chapter 2
SDGs in Village Development

Abstract Despite being the smallest unit of territorial subdivision in Indonesia, the village has complex socio-economic issues. If socio-economic problems in village areas can be resolved, then most of the development challenges in Indonesia will be resolved. This means the root of development problems in Indonesia can be identified in the village.

> ...if our villages really start to move forward on their own strength, then our entire society will also rise in rank and progress in all fields...
>
> - Sutan Sjahrir -

As one of the nation's founding fathers, Sutan Sjahrir wanted to emphasize that as a social entity, the village is an important marker for the progress of Indonesia. Therefore, if the Indonesian state and nation are to progress, the most important priority should be the progress of the village. That's because the villages are Indonesia, and Indonesia is the villages.

Despite being the smallest unit of territorial subdivision in Indonesia, the village has complex socio-economic issues. The village therefore serves as an evidence base for identifying various development problems (Prabowo 2013). If socio-economic problems in village areas can be resolved, then most of the development challenges in Indonesia will be resolved (Iskandar 2020). This means the root of development problems in Indonesia can be identified in the village. The nation's villages are a source of problem identification as they have the highest poverty levels, the largest proportion of the population with low health rates, low purchasing power, and low education levels compared to cities. For this reason, efforts to advance Indonesia through improvements to human development, the economy, and people's welfare should start with the village (Iskandar 2020).

In fact, in times of social and economic crisis, villages have proven to be the most resilient social entities. In other words, villages often have stronger socio-economic resilience than cities. Village administrative areas account for 91 percent of Indonesia's total land area (Ministry of Home Affairs 2019) and are home to 43 percent of the country's population, according to data from the 2020 National

A. H. Iskandar, *Accelerating the Achievement of the National Sustainable Development Goals in Indonesia*, Engaging Indonesia, https://doi.org/10.1007/978-981-97-8919-1_2

Census by Statistics Indonesia.[1] It therefore makes sense that Indonesia's development policies be centered on the villages. This is because the village is the strength of Indonesia, supporting the nation's socio-economic life and shaping its future.

2.1 Village Development in Indonesia

Since long ago, the village has played a strategic role as the smallest locus of development. In general terms, national development starts from the village. The village is identified as the biological mother of a country, which means the nuances of village life reflect the true picture of life in a country. Indonesia's strong value of mutual cooperation, stemming from the state ideology Pancasila, is a reflection of the behavior and social capital of the village community. This outlook has been described as the Indonesian horizon (Hermawan 2015).

The village is indeed a ubiquitous social entity. It is a common subject of discussions and conversations at all levels of society, from the lowest social bracket to the elite. Discourse about the village will never end, because the themes and topics that concern the village are so numerous and complex. Starting from economic and human potential, village issues also encompass poverty, underdevelopment, inequality, poor infrastructure, unemployment, and accessibility to health and education services.

Historically, the Indonesian village was born at the same time as the birth of the nation. In fact, prior to the independence proclamation of the Republic of Indonesia on August 17, 1945, there were more than 250 cultural entities,[2] which manifested into the thousands of villages that exist today. However, in the context of development and economy, from the time of ancient kingdoms and then the colonial era, the post-independence era and until the reform era, the position of the village has not been a priority (Amalik 2013). Many government policies have overlooked village communities (Mariana 2013). As an entity that existed before the Unitary State of the Republic of Indonesia, the village has an original structure and existence that cannot be contested. This means the state has an obligation to recognize the existence of the village both contextually and constitutionally.

This section describes the dynamics of the village in the Indonesian constitutional system, using a historical perspective. Soetardjo Kartohadikoesoemo (1984), one of Indonesia's founding fathers, has noted that in the colonial era, the regulations of the Dutch and Japanese colonial rulers recognized the village government system in juridical principles, as contained in Article 71 of the *Regerings Reglement* (the government regulation used as the basic law of the Dutch East Indies from 1854 to

---

[1] Statistics Indonesia is the official English name of Indonesia's central statistics agency, known locally as *Badan Pusat Statistik* (BPS).

[2] On the classifications of territory, the Elucidation of Indonesia's original 1945 Constitution states, 'Within the territory of the State of Indonesia there are approximately 250 "Zelfbesturende Landschappen [self-governing regions]" and Volksgemeenschappeu [village communities] such as villages in Java and Bali, nagari [states] in Minangkabau, hamlets and clans in Palembang, and so on'.

1926) and Article 128 of the *Indische Staatsregeling* (the constitution of the Dutch East Indies from 1926 to 1945) (Kartohadikoesoemo 1984). The same recognition was included in the Japanese colonial government's Regulation No. 1 of 1942 and Osamu Seirei Regulation No. 27 of 1942 (Soenardjo 1984). However, the reality is that most villages were controlled by the colonial bureaucratic system and its political economy (Syamsu 2008). In fact, during the Japanese colonial period, village autonomy was extremely limited; villagers were mobilized to meet Japan's wartime needs and became militia units such as *Heihol* (army assistants), *Keibodan* (police assistants), and *Seinendan* (youth front). In addition, village heads acted as supervisors for the cultivation of crops ordered by Japan, such as jatropha, rice, and sugar cane (Ministry of Home Affairs 2007). Such policies meant the village was positioned solely as an object of colonial exploitation.

In the early period of Indonesian independence, under founding President Soekarno, the government divided power between the center, the regions, and the villages by paying close attention to their conformity to the Constitution. There were four regional government laws that covered village autonomy, namely: (1) Law No. 22/1948 on Basic Principles of Regional Government with the Right to Regulate and Manage their Own Households; (2) Law No. 1/1957 on Principles of Regional Government; (3) Law No. 18/1965 on Principles of Regional Government; (4) Law No. 19/1965 on Civil Village Units as a Form of Transition to Accelerate the Establishment of Level III Regions throughout the Territory of the Republic of Indonesia. The village arrangements in the first two decades of independence are detailed in Table 2.1.

Indeed, considering the various policies related to villages at that time, it could be concluded that the government had laid a sufficiently strong foundation to position villages as autonomous regions. However, the unstable political dynamics of that era meant these laws did not work as effectively as intended.

After Old Order administration of Soekarno, the New Order government of President Suharto tended to increase the bureaucratization and centralization of development through the issuance of Law No. 5 of 1974 on *Pokok-Pokok Pemerintahan Daerah* (Principles of Regional Government). Despite its name, this law was by no means a decentralization-oriented policy to strengthen autonomy or form local governments; rather, it was oriented toward the formation of a central government that operates in the regions at a local state government level (Ministry of Home Affairs 2007). The issue of the village as a Level III Autonomous Region, which was previously discussed a lot, was no longer accommodated by the New Order government, especially after the enactment of Law No. 5 of 1979 on *Pemerintahan Desa* (Village Administration).

Law No. 5/1979 made the format of village administration uniform throughout Indonesia. One significant impact of this policy was the loss of the existence of 'special' or autonomous traditional villages. Article 1 of Law 5/1979 defines a village as follows:

> A village is an area occupied by a number of residents as a community unit, including as the legal community unit that has the lowest government organization directly under the district

Table 2.1 Comparison of village arrangements in early independence period

Law 22/1948	Law 1/1957	Law 18/1965	Law 19/1965
The territory of Indonesia is divided into three levels, namely province, regency (large city), and village (small town); The government recognizes the village as an agent of development to focus the orientation of development starting from the bottom up; The village is the lowest level autonomous region that is given the authority to regulate and manage its own households	The first regional autonomy law since the enactment of the semi-parliamentary government system and the 1950 Provisional Constitution; There are only two types of autonomous regions regulated by the government, namely Level I and Level II Regions; The formation of Level III Regions (district/municipality), must be prepared carefully because the village is the cornerstone of development	The government recognizes the village as a legal community entity with a unitary authority and the right to regulate and manage its own households; The establishment of Level III Autonomous Regions (district/ municipality) must pay attention to the lives of the people and their socio-economic progress, as well as applicable customary law; The regulation on villages remains part of this law, with the emphasis that there will be a law governing Level III Autonomous Regions	The government designated a new term to refer to the legal community unit, namely Civic Village (*Desapraja*); This law does not establish a new Civic Village, but recognizes existing legal entities with various names as Desapraja; Civic Village apparatus include the Village Head, Village Consultative Agency, Village Civil Service Officer, Village Registrar, Village Consultative Body; Civic Village is not the lowest hierarchy of regional autonomy, but it must in future be upgraded to a Level III Autonomous Region, as stipulated in Law 18/1965

Source Compiled from Academic Draft of Village Law, Ministry of Home Affairs, 2007

head and has the right to organize its own households within the bonds of the Unitary State of the Republic of Indonesia.

Thus, this Law strengthened the position of the central government to make the village the smallest government subordinate. Indirectly, this Law led to political corporatization efforts to encourage the creation of village political stability, strengthening bureaucratization and political control to build village loyalty to the government, and opening up space for capital penetration into villages by disbursing aid to villages in a manner that seemed populist (Eko 2005).

Hermawan (2015) made a similar observation, noting the village has experienced an economic-political crisis since the enactment of Law No. 5/1979, including: (1) The loss of village control over village land; (2) The destruction of the social basis of village autonomy, for example the village head tends to become subordinate to the district head and the regent; (3) The death of democratization because the village is managed in a centralized and authoritarian way; (4) The government takes advantage of rural poverty as a project commodity; (5) The inclusion of various aid programs into the village has further killed the village's independence or created dependence; (6) Massive exploitation and influx of capitalization make local resources limited; (7) Reduction of productive rural village communities (urbanization) because more production profits flow to the cities.

In addition, Law 5/1979 placed the village head not as a village community leader, but as a state functionary in a role that was more an extension of the bureaucracy. As a sole ruler, the village head tended to have full control over the livelihoods of all villagers and even held a concurrent position as head of the Village Consultative Council (Lembaga Musyawarah Desa/LMD), which was in charge of carrying out legislative functions. However, the village head had to submit and be accountable for all his or her activities to the regency, provincial, and central governments. There was no room for genuine democracy in the village decisions when the head was under the control and power of the central government.

The death of village democracy after the enactment of Law 5/1979 was seen not only in the changed role of the village head but also in the election of village heads. Although the village head was chosen by the village people through direct election, the fact is that such elections were manipulated and controlled by the supra-village powers, that is, the central and regional governments, through the formulation of political and administrative requirements. According to research conducted by Franz Husken (2001) in Pati of Central Java province, village head elections were rife with intimidation, manipulation of results and strictly controlled by the state.

After Suharto stepped down from the presidency in 1998, Indonesia's political system transitioned to a reform era, including a revival of democracy and decentralization for villages. The birth of the Law on Regional Government, namely Law No. 22 of 1999, as modified first by Law No. 32 of 2004, and then by Law No. 23 of 2014, was a manifestation of the government's commitment to the reform era to revive democracy, decentralization, regional autonomy, and villages as the most ideal and rational form of governance. The main points of village legislation under the three regional autonomy laws can be summarized as follows:

1. Law No. 22 of 1999
    a. The government recognizes the rights of origin, diversity, and uniqueness of the village as a self-governing community.
    b. The village is no longer placed as the lowest form of government under the district, but is a legal community unit that has the right to regulate and manage its own interests in accordance with the origin rights (*hak-hak asal usul desa*, literally 'the rights of origin') of the village.
    c. The village head is no longer the sole authority in the village, where previously the village head concurrently served as head of the Village Consultative Council.
    d. The institutionalization of village democracy is marked by replacing the Suharto era Village Consultative Council with the Village Council (Badan Perwakilan Desa/BPD), whose function is to protect customs, make village regulations, accommodate and channel people's aspirations, and supervise the implementation of village government.

e. Where previously the village head was responsible to the regent through the district head, this Law stipulates the accountability of the village head is given directly to representatives of the village community, namely through the Village Council.
f. There are at least three domains in which the power of the village head shifted to the Village Council, namely: (1) Decision-making in the form of village regulations is carried out jointly between the village head and the Village Council; (2) Financial management involves the Village Council, such as the preparation of the Village Budget (APBDes) and the auction of village treasury land; and (3) Recruitment, which used to be completely controlled by village heads, district heads, and regencies, was transferred to the Village Council.
g. The weakness of this Law lies in the unclear position of the village, because it confuses the principle of genuine autonomy (a self-governing community) with decentralization (local self-government). The Law completely surrenders village problems and arrangements to the regencies/municipalities, as if giving them a 'blank check' to determine village regulations. In the context of constitutional law, this system is quite contradictory, because it is the state that implements decentralization, not the regencies/municipalities (Ministry of Home Affairs, Academic Draft of Village Law, 2007).

2. Law no. 32 of 2004

a. In terms of position and rationale, there are no significant differences between Law No. 22 of 1999 and Law No. 32 of 2004. Both maintain the principles of diversity, participation, genuine autonomy, democratization, and community empowerment.
b. Village arrangements in this Law are more detailed than in Government Regulation No. 72 of 2005 on Villages.
c. There are five crucial issues in the substance and regulatory framework of this Law, which triggered debates and caused problems:
(1) The emergence of symptoms of recentralization can be seen in the relationship pattern between the central government and local governments, and the accountability of regional heads and village heads. This Law emphasizes a symmetrical power hierarchy from the center to the village, which means the upper level has greater control over the lower level. Article 200 of the Law states, 'In the Regency/Municipal Regional Government a Village Government is formed consisting of the Village Government and the Village Consultative Body'. This indirectly places the village as part of the regency government, even though the position of the village is autonomous. In addition, the accountability of the village head, which should be conveyed to the people as constituents, has actually been transferred to the regent/mayor through the district head.

(2) A stipulation that the village secretary must be appointed from the district-level civil service was criticized as a form of state bureaucratization that could cause problems for the village government. If the village secretary is a civil servant, then he or she has two-fold responsibilities and loyalties: to the village head and to the higher supervisory bureaucracy. In addition, jealousy arose among village officials, including the village head, on matters such as salary.

(3) There are still many regency/municipal governments that do not provide each village with a Village Fund Allocation (*Alokasi Dana Desa*/ADD). This is a development fund originating from the central government budget. The ADD was set at 10 percent of equalization funds received by regencies/municipalities and was to be distributed proportionally.

(4) This Law is silent on the management of village natural resources. It is more about the regulation of power, rather than the village government's function to ensure the welfare of its people through the regulation of sustainable management of natural resources, which are the livelihoods of villagers.

(5) Arrangements on inter-village cooperation and village cooperation with third parties are still not clearly discussed, such as what can be the basis and basic principles of inter-village cooperation, institutional formats, the amount of regency intervention in regulating inter-village cooperation, and how to model inter-village conflict resolution.

3. Law no. 23 of 2014

   a. The administrative division of state territory in this Law places the village as part of the district and under the state administration structure of the regional government.
   b. Village regulation in this Law is contained in only two articles, namely Article 371 and Article 372, and the structure of village government and its powers are not explained in more detail. This is different from the higher level of detail in Law No. 32 of 2004. Article 371 merely states the regulation of village authority is included in laws related to villages.

After Suharto's New Order era, efforts by successive governments to revive decentralization and village autonomy through these laws and regulations were unsatisfactory, especially for village associations (as village representatives). Debates, criticism, and demands to immediately formulate a shared vision and ideal arrangements for village autonomy intensified. In order to overcome fundamental problems concerning the essence, vision, and structural and institutional designs of villages, a political agreement was reached to make a separate Village Administration Law, outside the Regional Government Law. This agreement was the precursor to the birth of Law No. 6 of 2014 on Villages. The Village Law brought hope and a spirit of village renewal, as well as being critical of village-related laws and regulations that were considered insufficiently serious.

The enactment of the Village Law indicates the state has implemented resource redistribution and mandated authority and development to the nation's villages, preceded by full recognition and respect. This Law develops a new paradigm and concept of village governance, based on the principles of diversity and recognition of origin rights. It also allows for determination of local-level authority and decision-making. In addition, this Law uses two development approaches, namely 'Village Development' (where the village is the agent and subject of development) and 'Developing the Village' (where the agent of development is the government, acting with the village community). The Law also positions villages as inclusive, not exclusive, communities (Hermawan 2015).

In order to support the acceleration of village development, the Village Law strengthens two important points, namely the Village Fund Allocation sourced from the State Budget and other village income sources, such as village-sourced revenue, profit-sharing funds, grants, and financial assistance. The Village Fund Allocation, which had often been ignored by regency/municipal governments, is clearly regulated in Article 72 (Paragraph 4), which states the Village Fund Allocation is at least 10 percent of the equalization funds received by regencies/municipalities in the State Budget after deducting a Special Allocation Fund. The funds are expected to support capacity building and village development programs. In addition, villages are also given the opportunity to organize and manage their income sources independently. The Village Law not only strengthens the village economy, it also provides an opportunity to strengthen village democracy by affirming the division of power between the village head and the Village Council (Badan Permusyawaratan Desa/BPD). Under Law No. 32 of 2004, the village secretary had to come from the civil service, namely the district civil service. This requirement was abolished under the Village Law (Table 2.2).

2.2 Village Development in the Era of the Village Law

The Village Law changed the paradigm of village development, shifting the position from being only an object of development to now being the main subject of development. It has been over ten years since the Village Law was enacted on January 15, 2014, and of course the question is: How far has the Village Law changed the face of village development? In this section, the achievements of village development will be briefly but informatively reviewed, including a look at how villages contribute to national development.

Before discussing further the achievements of village development, we will first consider the condition of the village in 2015 by using a statistical approach using Developing Village Index (Indeks Desa Membangun/IDM). The IDM is prepared by the Ministry of Village, Development of Disadvantaged and Transmigration, by using three dimensions: Social, Economic, and Ecological resilience. The three measures are composed of 51 indicators that describe the availability and accessibility of services to village/rural communities. First, there were still many

Table 2.2 Comparison of post-1998 authorities held by villages

No.	Law	Authority
1	Law No. 22 of 1999	Authority already existing on the basis of the rights of origin of the village; Authorities which, under prevailing laws and regulations, have not been exercised by the regions and the government; Co-administrative 'assistance tasks' from the government, provincial government, and/or regency government
2	Law No. 32 of 2004	Existing administrative affairs based on the rights of origin of the village; Administrative affairs under the authority of the regency/municipality and whose arrangements are delegated to the village; Co-administrative 'assistance tasks' from the government, provincial government, and/or regency/municipal governments; Other affairs which are delegated by legislation to the village
3	Law No. 23 of 2014	Village authority not explained in detail; Only mentions the regulation of village authority will be included in laws related to villages
4	Law No. 6 of 2014	Authorities based on the rights of origin: Structuring the organizational system and institutions of indigenous peoples; Customary law institutions; Ownership of traditional rights; Management of customary village treasury lands; Management of customary land; Agreement in the life of the traditional village community; Filling the positions of village head and customary village apparatus; Term of office of the traditional village head Village-level local authorities: Village government sector; Village development; Village community; Village community empowerment Authorities assigned by the government, provincial government, or regency/municipal government; Other authorities assigned by the government, provincial government, or regency/municipal government in accordance with the provisions of prevailing legislation

Source Law No. 22/1999; Law No. 32/2004; Law No. 23/2014; Law No. 6/2014

villages with the status of Underdeveloped Village. Data from the Village Development Index (IDM) indicates the number of Underdeveloped Village in 2015 reached 13,453 (18.16%) out of total of villages in Indonesia. Most villages were classified as Disadvantaged Village with a total of 33,592 (45.34%) and Developing Village with a total of 22,822 (30.88%), while the remaining were classified as Developed (4.87%) and Self-Sufficient (0.23%) Villages. The issue of Underdeveloped Villages was a serious problem that demanded immediate resolution.

The IDM in Fig. 2.1 uses Village Potential data compiled by Ministry of Village, Development of Disadvantaged Regions and Transmigration, 2024.

Second, poverty in rural village areas was greater than in urban areas. Of the total 25.37 million poor people in Indonesia in 2024, 13.58 million (53.53%) lived

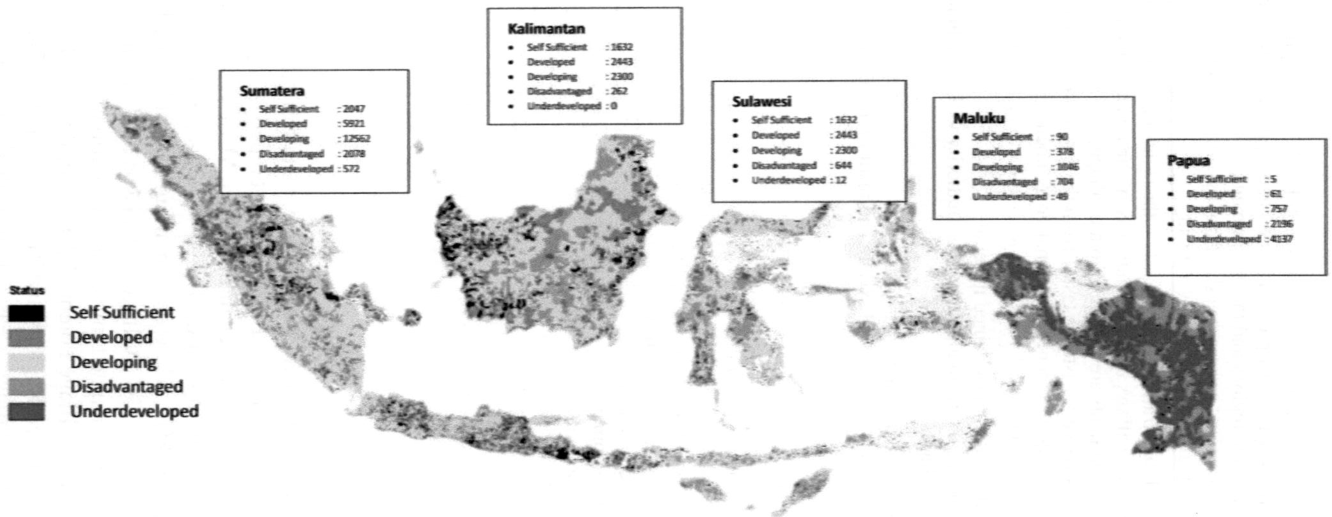

Fig. 2.1 Village development status on main islands, 2024. *Source* IDM Ministry of Village, Development of Disadvantaged Region, and Transmigration

in rural villages. Third, not all villages had received central government support for development and empowerment programs. According to Village Potential Data (2024), all villages throughout Indonesia were beneficiaries of community development and empowerment programs. Fourth, infrastructure and public services in the villages were still limited. Data on the need for village infrastructure development and services in 2024 is listed in Table 2.3.

Fifth, the quality of education in rural village communities was still low in 2014. Of the total 58.4 million people of productive age in the village, 60 percent had completed only elementary school (Statistics Indonesia 2020c). In the 1980s, about 78 percent of Indonesia's population lived in rural village areas. Now, more than half of Indonesia's people live in urban areas. The high rate of urbanization is a new problem and places a burden on urban areas if they lack sufficiently skilled residents. This trend of urbanization has also exacerbated the weaknesses of the village economy, which is characterized by low economies of scale, product quality that does not meet market specifications, and dominance by traders who bulk-buy crops in advance from farmers, many of whom have taken loans. These facts are the background of the village's economic performance, which was still at the bottom of the economic pyramid in 2014. In this context, it was necessary to complete and support economic policies starting from upstream to downstream sectors, with integration between sectors and stakeholders.

Under the leadership of President Joko Widodo, who won office in 2014, the government has been strongly committed to accelerate village development. This can be seen from the president's nine-point development agenda known as Nawacita, where the third component is aligned with suburban areas, especially villages, namely: 'Building Indonesia from the periphery by strengthening the regions and villages within the framework of the unitary state'. One of the concrete steps of

Table 2.3 Village infrastructure needs in 2024

No.	Infrastructure/service	Number of villages in need
1	Easy access to Village Health Post (Poskesdes), Integrated Services Post (Posyandu), Village Maternity Cottage (Polindes)	20.502
2	Drinking water	6.099
3	Public toilet	5.050
4	Village market	60.577
5	Asphalt-sealed roads	8.007
6	Mobile phone signal	23.657
7	Internet access	22.654
8	Banking institutions and Village Loan Agency	66.265
9	Public transportation	52.365
10	Road for vehicles of at least four wheels	15.514

Source Data and Information Center of the Ministry of Villages, Development of Disadvantaged Regions, and Transmigration the government's commitment is manifested in the distribution of funds under the Village Fund program.

The Village Fund program is not only the first of its kind in Indonesia, but is also the first and largest village funding policy in the world, covering 74,093 villages. When first disseminated in 2015, the amount of village funds allocated was IDR 20.67 trillion (USD 1.5 billion). Because village readiness was still rather low, only 82.72 percent of the village funds were absorbed at that time. However, this did not dampen the government's commitment and trust in village communities to continue to manage village funds. In 2023, the village fund was more than tripled to IDR 70 trillion. Based on this trust, the absorption of village funds increased from 82.72 percent to 99.80 percent. Witnessing the progress made by the villages in managing the village funds, the government did not hesitate to increase the allocation of village funds again in subsequent years (Table 2.4).

In the five-year period from 2015 to 2023, through the provision of the Village Fund, village communities have succeeded in building infrastructure to meet their basic needs, both to support economic activity and quality of life. Infrastructure-supporting economic activities include 350,755 km of village roads; 1904 km of bridges; 14,612 village markets; 43,123 village-owned enterprise activities; 9232 jetties; 6707 tendons; 539,938 land retainers; and 603,348 irrigation units. Meanwhile, the supporting infrastructure for meeting basic needs includes 1.742.886 clean water facilities; 536,907 units of bathing, washing and latrine facilities; 33,897 sport facilities; 45,993 integrated services posts; 52,940,731 m drainage; and 70,644 early childhood education units. Such massive village infrastructure development has never before happened in Indonesia's history. The success is proof that if given both trust

Table 2.4 Distribution of village funds for 2015–2024

No.	Year	Allocation (trillion rupiah)	Villages	Average per village (million rupiah)	%Transferred from cash account		
					State to Region (%)	Region to village (%)	State to village (%)
1	2015	20.67	74,093	280.3	82.72	82.72	
2	2016	46.98	74,754	643.6	97.65	97.65	
3	2017	60	74,910	800.4	98.54	98.54	
4	2018	60	74,957	800.4	99.62	99.62	
5	2019	70	74,954	933.9	98.52	98.52	
6	2020	71	74,954	949.78			99.95
7	2021	72	74,961	960.5			99.80
8	2022	68	74,960	907.13			99.86
9	2023	70	74,954	933.9			99.80
10	2024	71	75,259	943.33			TBC

Source Village Development Information System, Ministry of Villages, Development of Disadvantaged Regions, and Transmigration

and support, the village apparatus and community will gradually be able to overcome the problem of limited village infrastructure.

Although the distribution and use of village funds has been quite effective, it is acknowledged that there are still some problems regarding the use of the funds. There are some village officials and other elements who deliberately misuse village funds. Indonesia Corruption Watch recorded 298 cases of corruption in village funds over 2015–2023. This number is actually quite small when compared to the amount of village funds that were used appropriately. The relatively low level of fund misuse is inseparable from the government's efforts to encourage collaboration between parties to carry out supervision, including by universities, the mass media, religious organizations, and society in general.

The impact of village funds is felt not only through the ongoing massive development of village infrastructure, but even more through the achievement of higher incomes, lower poverty, higher employment, and lower economic inequality. For example, the Village Cash-for-Work scheme is sourced from the Village Fund. From 2015 to 2023, the Cash-for-Work scheme succeeded in providing 120,453,554 days of work and absorbed a workforce of 14,666,026 people through a total of 483,165 activities. In addition to providing jobs for local workers, the Cash-for-Work projects also increase the earnings of village businesses, because most of the building materials purchased for the projects come from local shops. A 2018 study by the Ministry of Agriculture's Center for Agricultural Economics and Policy found that village funds are able to increase the production and income of farmers who benefit indirectly from the construction of village roads, farm roads, irrigation networks, drainage systems, and retaining walls, as well as agricultural training.

A study conducted in 2019 by 33 universities grouped in the Higher Education Forum for Villages listed several benefits generated by village funds in 33 provinces (Ministry of Villages, Development of Disadvantaged Regions, and Transmigration, 2019). Some of the key benefits are as follows:

a. Development of village facilities and infrastructure, resulting in: (1) smooth running of social activities (traditional community activities, youth activities, reading gardens, early childhood education units, easy access to elementary school/junior high/senior high); (2) opening of the community's economic access, job opportunities, and higher incomes.
b. Self-management of building and infrastructure development, resulting in: (1) infrastructure development is considered to be of good quality; (2) increasing involvement of local workers; and (3) increasing circulation of money in the community.
c. Community empowerment, resulting in: (1) greater role of residents in village deliberations and (2) more proposals for community activities are approved as village development programs.
d. Improvement of community capacity, resulting in: (1) higher skills in building infrastructure, farming, agro-industry processing; (2) improved organizational

experience; and (3) improved relations with government institutions for management of village-owned enterprises, with marketing institutions for agricultural products, and with financial institutions for strengthening village capital.

e. Increased agricultural productivity, resulting in: (1) smooth running of farming businesses due to widespread construction of village roads and farm roads, reservoirs, bridges, irrigation systems; (2) capacity building, skills, and changing attitudes to be more open to agricultural innovation, leading to empowerment of farmers, farmers' groups, village-owned enterprises, and cooperatives (providing seeds, agricultural crops, fertilizers).
f. Construction of village roads improves transportation by: (1) facilitating the flow of transport from outside the village to the village; (2) increasing the different types of vehicles that can access the roads, especially diesel trucks and pickup trucks; and (3) decreasing travel time to the district capital.
g. Economic access, resulting in: (1) more money circulating in the community; (2) more types of goods traded, especially agricultural products; and (3) smoother purchases of merchandise, agricultural production tools, and household goods.
h. Health benefits, namely: (1) improved community access to clean water and (2) significant reductions in diarrheal diseases and skin diseases.
i. Education, including: easier public access to schools (early childhood education, elementary school/junior high school/senior high, equivalency education) with the construction of roads and bridges, and provision of modes of transportation.
j. Household social status mobility, namely: (1) the perception of increased household social status, in which households previously perceived as poor in 2014, have now increased in social status; (2) households are more respected and more involved in village development.

The village funds have clearly benefited various aspects of village life, ranging from economic, social, educational, and health to the level of community participation in village planning and development.

Since 2014, the government has been relatively successful in improving village welfare. This success is due to the effective performance of village funds, supported by other government programs, amid continuing improvement of transparency, participation, and collaboration. Such success is evidenced by increases in income of rural village communities, while poverty and unemployment have also been reduced, and economic inequality is kept low. The income per capita of villagers increased during 2014–2023 with an average increase of 7.8 percent. In 2023, the average monthly income of villagers reached IDR 1,119,119, higher than the 2014 income level of IDR 572,586 (Fig. 2.2).

Regarding rural village poverty, from 2015 to 2024 there was an average decrease of 0.26 percent. In 2024, Indonesia's poverty rate was in single digits, its lowest rate since 1998. The number of poor people declined by 3.34 million, of whom 1.29 million (62.4%) were in rural areas. (BPS 2024). If you look at Statistics Indonesia data, poverty is still higher in rural areas than in urban areas. However, the gradual decline in poverty in rural areas has been higher than in urban areas. An increase

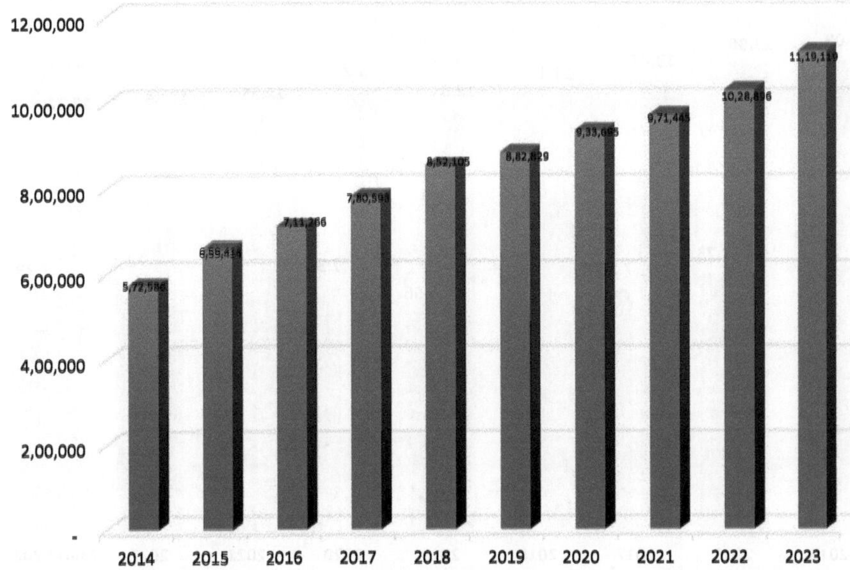

Fig. 2.2 Monthly income per capita in village communities, 2014–2023 (IDR). *Source* BPS

in poverty in March 2020, due to the policy of Large-Scale Social Restrictions to prevent the spread of COVID-19, was also higher in cities (Fig. 2.3).

The percentage of poverty reduction in rural village areas and urban areas is shown in Table 2.5.

The village economy has grown with the development of village-owned enterprises (known locally as *Badan Usaha Milik Desa*, BUM Desa). Since 2014, the number of BUM Desa has increased sharply. At the end of 2014, the number of BUM Desa reached only 1022, but by mid-2024 it had soared to 65,941. However, the development of BUM Desa still faces several problems such as limited capital and investment, low market share, weak operational management, low human resources competence, sub-optimal technology, and potential for moral hazard. Despite these challenges, several BUM Desa have become the main pillars of village economic activities. Data from the Ministry of Villages, Development of Disadvantaged Regions, and Transmigration (2019) indicates 584 village-owned enterprises have an annual turnover of more than IDR 1 billion, 465 have a turnover of between IDR 500 million and 1 billion, and 3212 have a turnover of IDR 100 million to IDR 500 million (Fig. 2.4).

The increasingly equitable development of village infrastructure, as well as the emergence of new jobs, higher community incomes, and rising village economies are all in line with the increase in the Village Development Index (IDM). According to IDM 2024, the condition of the village status is statistically improved. The self-sufficient status is increased significantly, with total of 11,456 (15.2%).

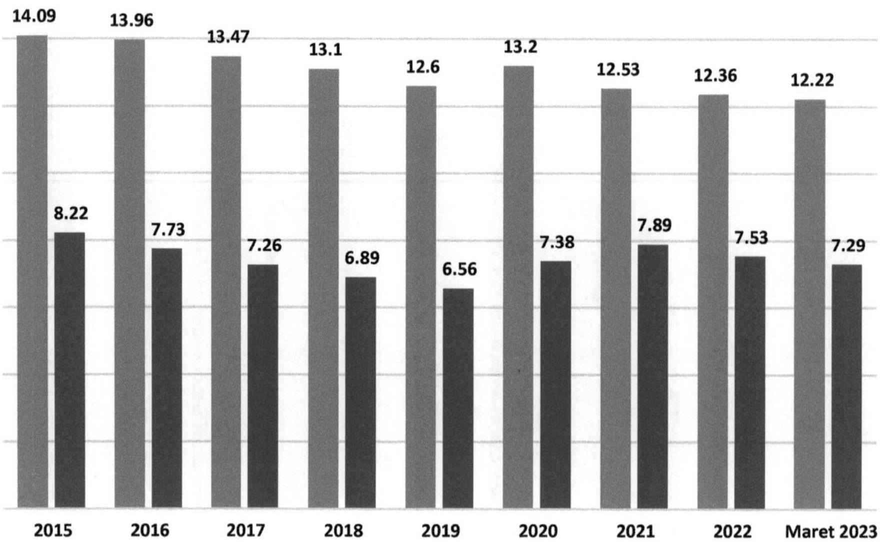

Fig. 2.3 Rural village and urban poverty rates, 2015–2020 (%). *Source* BPS

Table 2.5 Decreases in rural and urban poverty, 2016–2020

Year	Decrease	
	Urban (%)	Rural (%)
2016	– 0.49	– 0.13
2017	– 0.47	– 0.49
2018	– 0.37	– 0.37
2019	– 0.33	– 0.5
2020	0.82	0.6
2021	0.51	– 0.67
2022	– 0.36	– 0.17
2023	– 0.24	– 0.14
2024*	– 0.2	– 0.43

Source BPS

Underdeveloped village decreased to 4850 (6.44%) and Disadvantaged decreased to 7154 (9.51%) out of total villages in Indonesia. Most villages now are in Developing (38.22%) and Developed (30.6%) status (Fig. 2.5).

This achievement exceeded the targets of the 2020–2024 National Medium-Term Development Plan (RPJMN), namely: (1) Raising the performance of 5000 villages to become Self-Sufficient Villages; (2) Assisting 5000 villages to become Developing Villages; (3) Decreasing 10,000 disadvantaged villages.

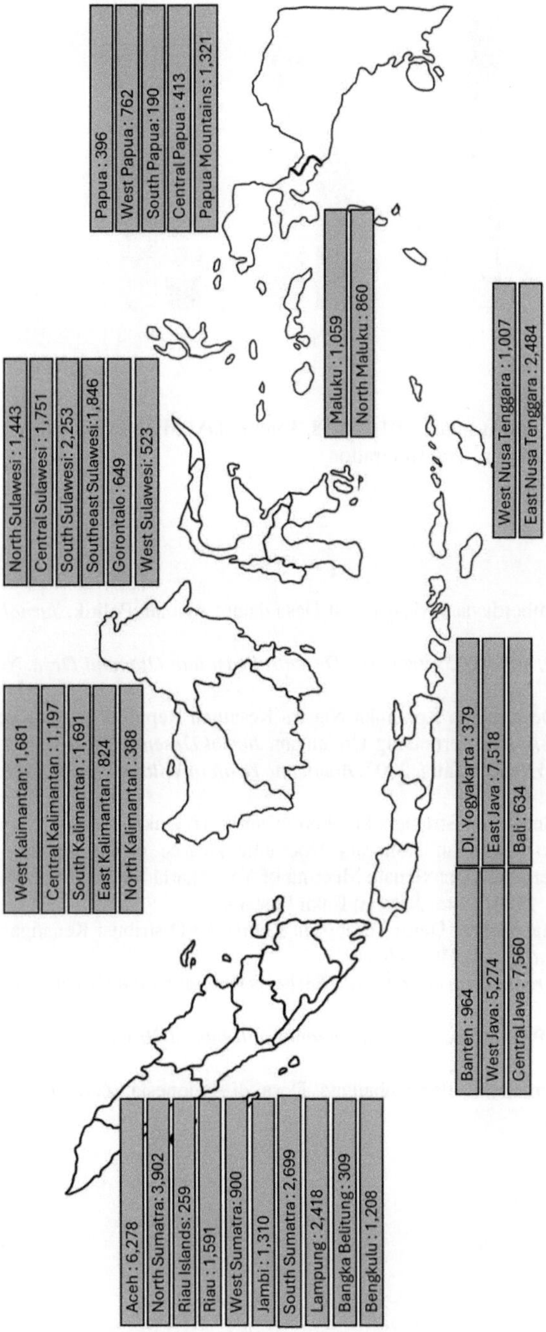

Fig. 2.4 Village-owned enterprises per province, 2024. *Source* Ministry of Villages, Development of Disadvantaged Regions, and Transmigration

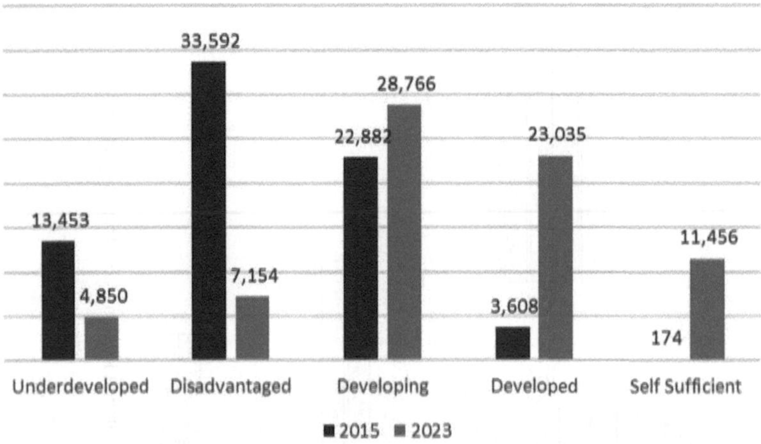

Fig. 2.5 Village development status, 2014–2018. *Source* IDM Ministry of Villages, Development of Disadvantaged Regions, and Transmigration

References

Amalik, M.J. 2013. Pemberdayaan Masyarakat Desa dalam Ayunan Politik. *Jurnal Mandatory* 10 (1).

Eko, S., and A. Rozaki, eds. 2005. *Prakarsa Desentralisasi dan Otonomi Desa*. Yogyakarta: IRE Press.

Hermawan, R. 2015. Desa dalam Kerangka Negara Kesatuan Republik Indonesia: Melihat Desa dari Sudut Pandang Aturan Perundang-Undangan. *Jurnal Desentralisasi* 13 (1): 61–75.

Indonesian Ministry of Home Affairs. 2007. *Academic Draft of Village Law*. Jakarta: Home Affairs Ministry.

Iskandar, A.H. 2020. Kebijakan Strategis Pemberdayaan Masyarakat Menuju Kemandirian Desa. Scientific speech to obtain an Honorary Doctorate Degree in Community Empowerment Management, delivered at Open Senate Meeting of Yogyakarta State University.

Kartohadikoesoemo, S. 1984. *Desa*. Jakarta: Balai Pustaka.

Mariana, D. 2013. Dana Alokasi Desa: Terobosan Kebijakan Distribusi Keuangan Negara untuk Desa. *Jurnal Mandatory* 10 (1): 52–62.

Prabowo, T. A. 2013. *Perencanaan Desa Wisata Berbasis Pembangunan Berkelanjutan*. Yogyakarta: Dinadra Creative.

Statistics Indonesia (BPS). 2020c. *Statistik Indonesia Tahun 2020* (Indonesian Statistics 2020). Jakarta: BPS.

Syamsu, S. 2008. Memahami Perkembangan Desa di Indonesia. *Government: Jurnal Ilmu Pemerintahan*. 1 (1): 77–88.

Chapter 3
The Urgency of the SDGs Desa

Abstract After centuries of colonial occupation and exploitation, the proclamation of Indonesian independence on August 17, 1945, brought the nation's people to the gateway of fields of prosperity. For the village, the fields of prosperity are now formulated in the development concept within Law No. 6/2014 on Villages.

Heavy, we carry together; light, we carry together.
Indonesian proverb

After centuries of colonial occupation and exploitation, the proclamation of Indonesian independence on August 17, 1945, brought the nation's people to the gateway of fields of prosperity. For the village, the fields of prosperity are now formulated in the development concept within Law No. 6/2014 on Villages.

Even though development is a large, monolithic narrative, it must be applied according to unique local sociological conditions (Alatas 2010; Heryanto 1990; Thomas 2002). It is important to have a distinctive meaning of development for Indonesian villages, so that all residents can easily participate at all levels, from planning to implementation and obtaining results (Sajogyo 2006). The formulation of concepts, measurements, and management that best suit the lives of residents are useful in illuminating the development paths for fulfilling the fields of prosperity.

3.1 Three Levels of Village Development

Article 78 of Law No. 6/2014 on Villages emphasizes the concept of three levels of development: norms, goals, and methods. Norms occupy the most abstract position, goals refer to the expected new structure, while methods describe strategies, mechanisms, or procedures to achieve the goals and norms.

Paragraph 3 of the Law focuses on the abstraction of development norms, namely togetherness, kinship, mutual cooperation, mainstreaming peace, and social justice. These norms mean that village development should involve as much community

A. H. Iskandar, *Accelerating the Achievement of the National Sustainable Development Goals in Indonesia*, Engaging Indonesia, https://doi.org/10.1007/978-981-97-8919-1_3

participation as possible, jointly planning, implementing, and reaping the results (Sajogyo 2006). At the same time, development risks are also shared, with communities working together to deal with problems, or by applying local knowledge. This concept of sharing both the benefits and burdens is reflected in an Indonesian proverb, which states: Heavy, we carry together; light, we carry together (Mubyarto 1996).

Village development is not neutral (Esteva 1992), so development measures must clearly indicate which parties' interests should be prioritized. Social justice norms tend to defend the lowest strata of society (Mubyarto 1987; Sajogyo 1997a, b).

Development inevitably involves change (McMichael 2003), which is often fundamental to the village. However, it is hoped that throughout the development process, harmonious relations between residents can be maintained. Such harmony is included in the norm of peace by avoiding conflict as much as possible during the village development process (Kartohadikoesoemo 1984).

Paragraph 1 of Law No. 6/2014 lays out the goals of development, namely welfare, quality of human life, and poverty alleviation. Welfare covers a broad spectrum, not infrequently encapsulating all development goals. When defined narrowly, welfare is closer to prosperity, meaning that villagers' needs are being fulfilled, even leading to surpluses, which can then be exchanged with other parties.

Quality of human life is commonly understood as a component of the United Nations' Human Development Index (HDI), which covers education, health, and income (UNDP 1990). Examined more deeply, an important sub-component is gender equality.

Poverty reduction is perhaps the most familiar goal of village development. That's because villages are often viewed as being synonymous with underdevelopment and poverty, whereas village development in Indonesia has always been directed toward overcoming poverty. The study of village poverty has been widespread, so its characteristics are well known (Agusta 2014; Sajogyo 2006). The issue of poverty is often focused on the accuracy of data on the poor, as well as the concept of poverty as an indicator of underdevelopment in the village.

Paragraph 1 of Law No. 6/2014 also outlines strategies, mechanisms, or ways to achieve village development goals, namely through fulfilling basic needs, building facilities and infrastructure, developing local economic potential, and utilizing natural resources and the environment in a sustainable manner. Since the 1970s, development oriented toward the fulfillment of basic needs refers to the provision of employment or escape from unemployment, decent residential infrastructure to support work and life, as well as basic education and health facilities (Friedmann 1992).

The development of facilities and infrastructure can be the most obvious need for the village. In fact, the village can often be differentiated from the city because the village lacks such facilities and infrastructure (Agusta and Soleh 2004). Therefore, the formulation of the objectives of infrastructure development is easy to match with the needs felt by the villagers themselves.

Development enables entry into the economy of exchange, especially through the medium of money. In order to excel in these exchanges, villages need to discover their economic potential so they can obtain added value. Economic development cannot be limited by area, but the village's focus on profit potential can strengthen its bargaining position when dealing with suppliers of business raw materials, as well as buyers of the village's goods and services. Such a mutually supportive local economic ecosystem can provide positive benefits for villagers (Blakely and Bradshaw 2002).

Natural resources need to be utilized for development, while considering the sustainability of benefits for future generations. In the initial stage of development, the challenge is to ensure that natural resources available in the village can be utilized for the welfare of the residents themselves. The next stage must involve the search for technologies that can process resources more efficiently, even those capable of renewing natural resources that have been processed.

3.2 Disseminating the Concept of Village Development

Article 78 of Law 6/2014 on Villages provides ways to achieve all norms, goals, and strategies of village development. Its components include planning, implementation, results, benefits, and impacts (Kroupova et al. 2009). These can be added to the underlying vision or play a role in mainstreaming development.

If applied to this arrangement, the previously mentioned village development norms occupy the position of a vision that must become the central focus of all efforts to disseminate the concept of development. The long-term vision of village development is the creation and maintenance of togetherness, kinship, mutual cooperation, mainstreaming peace, and social justice.

These development objectives occupy the position of expected development impacts. Thus, the objectives of village development are formulated as the achievement of welfare, quality of human life, and poverty alleviation of villagers.

Meanwhile, the development strategy occupies the position of development implementation, acquisition of results, and growth of expected benefits. Results are measured by aspects of meeting basic needs, as well as the development of facilities and infrastructure. The expected benefits are manifested by development of local economic potential, as well as the sustainable use of natural resources and the environment.

Village development plans are formulated through the six-year Village Medium-Term Development Plan (RPJMDes), the annual Village Government Work Plan (RKPDes), and the annual Village Budget (APBDes). Through these plans and budgets, the villages have felt the breadth of the scope of village development. The scope of activities in the Village Government Work Plan and the Village Budget are comprehensive, as are the targets for achieving development benefits and impacts. In each village, the activities cover all aspects of society, including governance, development, community empowerment, community building, and disasters. In the daily life of village development, all the components of development feel interconnected and difficult to separate.

As an illustration, in 2019 alone there were 9 million types of activities included in the village budgets, an increase from 7.5 million in the previous year. Thus, the challenge that arises from the field is to incorporate village aspirations into concepts of an adequate size. This is because, so far, the concepts and measures of development have focused only on certain aspects of the village. Thus, village progress will only be partial.

In fact, efforts to expand the reach of development by increasing the discussion from results (output) to benefits (outcomes) and impacts can be hindered by sectoral restrictions. For example, narrowing the focus of poverty reduction efforts can indeed increase the income of the poor and reduce the economic lag of the village, but at the same time it ignores village progress. Less attention is paid to accelerating the development of the village's middle class, while the wealthiest class in the village is less directed toward supporting the economic progress of the middle and lower classes. This means it is difficult for poverty alleviation measures to detect the acceleration of prosperity outside the poor.

3.3 Disadvantages of Measurements for Village Development

Development is more practical when viewed through the measures created to build village narratives. Some of these measures are classified in a specific concept. For example, the Fire Aware Self-Sufficient Village (*Desa Mandiri Peduli Api*, DMPA) is focused on the economic activities of residents, as well as activities for preventing and handling land fires around the village. This measure was chosen from among economic empowerment activities, especially in agriculture, fire-fighting training, and land management, so that villagers could prevent land fires or respond to fires.

Similarly, the Peat Care Village (*Desa Peduli Gambut*, DPG) guides village progress by increasing knowledge about peatlands, peat-specific land management, and reducing peatland fires. As another example, the Inclusive Village concept directs allocations from the village budget to support business facilities and daily activities for persons with disabilities. Tourism Villages concentrate village spending on the construction of tourist facilities and establish Village-Owned Enterprises in the tourism sector. Productive Migrant Villages focus on development for both male and female migrant workers. Digital Villages spend on information technology facilities and signal processing infrastructure for security, economy, ecology, and so on.

Such specific emphasis on certain aspects of village development has the advantage of focusing budgets and other resources on certain activities. The results are positive for residents who make a living from that particular aspect. For example, a pottery craftsperson benefits from tourism villages. However, a specific development focus may not deliver benefits to residents who make a living outside the focus area. Coastal Tourism Villages, for example, do not always provide direct benefits for those involved in agriculture or industry.

Efforts are made to measure as many aspects of development as possible, in a composite index, for example, in the Developing Villages Index (IDM) and the

Village Development Index (IPD). Foreign institutions such as the World Bank, Organization for Economic Cooperation and Development (OECD), and Food and Agriculture Organization (FAO) also compile a series of components into composite indices (United Nations 2007).

Nevertheless, all these composite measurements have not yet succeeded in covering all aspects of village development. The scope of the FAO is closely related to agriculture, so it is insufficient to characterize village industry and tourism. Measures from the OECD and the World Bank often emphasize the existence of forests and green open spaces. This is suitable for Forest Villages, but lowers the status of villages around urban areas.

Indonesia presently uses two measures of village development, the aforementioned Developing Villages Index (IDM) and Village Development Index (IPD). The IDM uses social, economic, and environmental dimensions. The environmental weight is much higher than the social aspect, and the lowest weight is in the economic aspect. This composite index does not pay attention to aspects of village governance. Consequently, Self-Sufficient Villages tend to be found in locations that are not polluted, do not experience disasters, and care for the environment. Such villages are located far from urban areas.

The IPD focuses on measuring the fulfillment of minimum services in the village (National Development Planning Agency and Statistics Indonesia, 2015). Minimum services include basic services, infrastructure conditions, accessibility and transportation, public services, and governance. The advantage of the IPD is in measuring the extent to which the minimum needs of the community have been met. Unlike the IDM, there is no environmental dimension in the IPD. In addition, the IPD's orientation toward minimum needs limits its measurement to Disadvantaged Villages and those improving from Disadvantaged status. It is therefore difficult for the IPD to measure the progress of Self-Sufficient Villages.

Weaknesses in the composite measurements are also due to the types of data used. The IPD and IDM both use regional data from the village level, not from the household or individual level. The IDM collects data on villages from annual civic records registered by village heads. The IPD is based on village potential data from Statistics Indonesia. As the village is seen as a single entity, this kind of data is more useful for the government authorities above the village and is less useful for the village itself. The central and regional governments can distinguish one village from another. However, for the village itself, it is not possible to distinguish the condition of one neighborhood unit from another. For example, there may be data on asphalt-sealed roads in a village area; however, it is not known in which specific neighborhood unit the roads are sealed, while some units may have unsealed roads.

The wording of questionnaires also affects data and the calculation of composite indices. The collection of village territorial data is limited to the presence or absence of basic facilities such as elementary schools, junior high schools, senior high schools, integrated services posts, village maternity cottages, and the like. The data does not include the quality of such facilities, let alone their level of public service. Due to the bias of this kind of available information, the result is a measurement that is limited to a bare minimum: the presence or absence of basic amenities. This measurement is useful for detecting the extent to which village development is able to eradicate Disadvantaged and Very Disadvantaged Villages. For example, it is easy to distinguish Disadvantaged Villages that have become Developed Villages in disadvantaged, border, and outlying areas. However, in villages where minimum needs have been met, it is difficult to distinguish further progress. That's because there is no further measure of progress. For example, further differences between Developed Villages and Self-Sufficient Villages in Java and Bali are difficult to distinguish because there are no measures and data for the state of village progress (Fig. 3.1).

Highest average IDM scores were achieved by Southern Papua, Central Papua, Mountains Papua, West Kalimantan, and Riau. This indicates the acceleration of village development is higher outside Java, increasing rural and urban equity more quickly.

It has been proven that development programs are more effective when they directly target families or individuals (United Nations 2007). Moreover, it is easier to measure success when the beneficiaries of a program are known by name and by address. The problem is that this kind of data is rare, being gathered only every 10 years when a national census is conducted. The compilation of such data is also quite expensive with the current method of national census collection.

However, it is clear that village development will be more beneficial and effective when directly focused on individual villagers, one after another. At the same time, measures that are more focused on improving the quality of individuals or families are also needed.

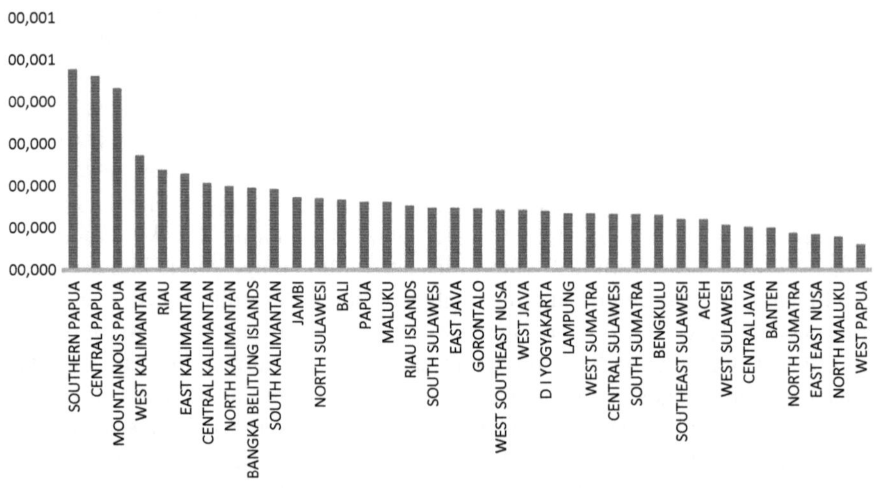

Fig. 3.1 Developing villages index, 2015–2023. *Source* Ministry of Villages, Development of Disadvantaged Regions, and Transmigration

3.4 Position of the SDGs

In short, development policies need to be more in line with actual conditions in the field. Development policy directions must be detailed, covering methods, substance, and final goals, and should be focused across very small areas to maximize benefits to families and individuals. Devising such direction is not easy, because it starts with the most comprehensive concept of development, followed by the formulation of questions, covering both the availability and quality of facilities, as well as censuses at the family and individual levels.

The Sustainable Development Goals (SDGs), known in Indonesia as *Tujuàn Pembangunan Berkelanjutan* (TPB), are positioned to fill all these needs. The SDGs comprise the main development concepts compiled by the United Nations: social development, economic development, environmental development, resilience to disasters, population dynamics, and so on. The SDGs fulfill the need for development that is more in line with real conditions on the ground. The SDGs consist of 17 goals, 169 targets, and 241 indicators. At the national level in Indonesia, under Presidential Regulation 59 of 2017, implementation of the SDGs has swelled to 319 indicators within the scope of the 17 goals (Table 3.1).

The meaning of sustainability in development is shown primarily by objectives related to conservation of the natural environment, such as using clean and renewable energy, tackling climate change, protecting marine and terrestrial ecosystems, and building sustainable cities and communities. Sustainability is also demonstrated by improving the quality of economic growth, taking into account the improvement in the quality of job seekers so they can obtain decent work. Additional indicators include innovations in industrial and infrastructure activities, environmentally friendly production, and responsible consumption.

It can be said the village occupies a 'favorable' position when it comes to environment-related goals. That's because the natural environment in the village is relatively more preserved in the absence of significant exploitation by humans. This is an important positive point, even before starting village development.

A distinctive feature of the SDGs is that they are formulated in a statement of objectives, with indicators used to measure progress toward each target. People are at the heart of the SDGs, with the aim of leaving no one behind, meaning the target is always fulfillment for 100 percent of the population. This focus on people is contained in SDGs 1 to 6: no poverty, zero hunger, good health and well-being, quality education, gender equality, and clean water and sanitation.

As already mentioned, the objectives related to physical areas are directed at indicators concerning availability and sustainability (taking into account long-term or sustainable benefits) and managing their impact on local ecosystems.

The goals themselves not only contain the substance of development but also encompass strategies to achieve them, including the required institutions. Climate and ecosystem management does not recognize community boundaries or official territories, so cooperation between parties is essential. In line with increasing knowledge on the variety and nature of stakeholders, cooperation involves at least the

Table 3.1 Conformity of the SDGs with village development under Law 6/2014

Level	Law 6/2014 on villages	SDGs
Village development Norms	Togetherness, kinship, social justice, mainstreaming peace, and mutual cooperation	SDG 5: Gender Equality SDG 10: Reduced Inequality SDG 16: Peace, Justice, and Strong institutions SDG 17: Partnerships for the Goals
Village development Goals	Poverty alleviation, welfare, quality of human life	SDG 1: No Poverty SDG 2: Zero Hunger SDG 3: Good Health and Well-being SDG 4: Quality Education
Village development Method	Fulfillment of basic needs, development of facilities and infrastructure, development of local economic potential, and sustainable use of natural resources and the environment	SDG 6: Clean Water and Sanitation SDG 7: Affordable and Clean Energy SDG 8: Decent Work and Economic Growth SDG 9: Industry, Innovation, and Infrastructure SDG 11: Sustainable Cities and Communities SDG 12: Responsible Consumption and Production SDG 13: Climate Action SDG 14: Life Below Water SDG 15: Life on Land

community, the state, and private parties. The SDGs are indicated by the goals of peace, justice, and strong institutions, as well as partnerships to achieve these goals.

3.5 The Urgency of the SDGs Desa

Each village covers a relatively small working area with a small number of residents. This facilitates the management of matters such as settlements, infrastructure, and land and sea ecosystems. It also facilitates the handling of the population in terms of poverty, hunger, health, education, and social justice among citizens.

Although small in area, each village is part of the national bureaucracy, its people are part of the Indonesian citizenry, and its ecosystem is part of the wider ecosystem, so the substance of development in the village is also complex.

It is in this context that the SDGs need to be implemented at the village level. Namely, developing each village with the complete substance of the SDGs, yet in a controlled manner within a limited area and among local residents. This maximizes the likelihood of successful implementation of the SDGs in each village. This is a boon to efforts to implement the SDGs in the field.

As the SDGs have become part of the global development dialogue, their implementation in the villages puts Indonesian villagers among the global citizenry. Without much fuss, the SDGs were included in the Regulation of the Minister of Villages, Development of Disadvantaged Regions, and Transmigration No. 13 of 2020 on Priorities for Use of Village Funds in 2021.

The Regulation was positioned at the program level, which is more abstract, to formulate a new perspective on village development. In 2021, village development was directed toward reducing poverty, maintaining health, achieving inclusiveness for women and disabled groups, meeting energy needs, creating job opportunities and economic growth, as well as responsible consumption and production.

Such annual regulations are used as the basis for the preparation of development activities and budgeting in each village. The conditions, potentials, and problems of each village are compiled at the beginning of the year and documented for the public through the website of the Ministry of Villages, Development of Disadvantaged Regions, and Transmigration. This data is used by the village as a starting point for the preparation of activities and budgeting, which are determined by the Village Council. Thus, the accomplishment of development activities also accelerates the fulfillment of the SDGs.

References

Agusta, I. 2014. *Diskursus, Kekuasaan, dan Praktik Kemiskinan di Pedesaan*. Jakarta: Yayasan Obor Indonesia.

Agusta, I., and A.M. Soleh. 2004. *Jejak-jejak Kesejahteraan: Evaluasi Benefit Program Pembangunan Prasarana Pendukung Desa Tertinggal*. Jakarta: Binasiamindo Aneka.

Alatas, S. F. 2010. *Diskursus Alternatif dalam Ilmu Sosial Asia: Tanggapan terhadap Eurosentrisme*. Indonesian translation. Bandung: Mizan Media Utama.

Blakely, E. J., and T. K. Bradshaw. 2002. *Planning Local Economic Development: Theory and Practice*. Thousand Oaks, California: SAGE.

Esteva, G. 1992. Development. In *The Development Dictionary: A Guide to Knowledge as Power*, ed. W. Sachs. London: Zed Books.

Friedmann, J. 1992. *Empowerment: The Politics of Alternative Development*. Cambridge: Blackwell.

Heryanto, A. 1990. The Making of Language: Developmentalism in Indonesia. In *Prisma: The Indonesian Indicator*, No. 50.

Kartohadikoesoemo, S. 1984. *Desa*. Jakarta: Balai Pustaka.

Kroupova, Z., G. Cervena, and M. Antouskova. 2009. Tools for Agriculture and Rural Development Policy Evaluation. *The Business Review* 14: 1.

McMichael, P. 2003. *Development and Social Change: A Global Perspective*. Thousand Oaks: Pine Forge Press.

Mubyarto. 1987. *Ekonomi Pancasila: Gagasan dan Kemungkinan*. Jakarta: LP3ES.

Mubyarto. 1996. *Menuju Gerakan Nasional Penanggulangan Kemiskinan*. Yogyakarta: Aditya Media.

Sajogyo. 1977a. Golongan Miskin dan Partisipasinya dalam Pembangunan Desa. In *Prisma*, vol. 6, No. 3.

Sajogyo. 1997b. *Menciptakan Visi Mendukung Pengembangan Kelompok Swadaya Mandiri dalam Gerakan Nasional PPK*. Jakarta: Puspa Swara.

Sajogyo. 2006. *Ekososiologi: Deideologisasi Teori, Restrukturisasi Aksi (Petani dan Perdesaan sebagai Kasus Uji)*. Yogyakarta: Cindelaras Pustaka Rakyat Cerdas.

Thomas, A. 2002. Meanings and Views of Development. In *Poverty and Development: Into the 21st Century*, ed. T. Allen, and A. Thomas. Oxford: Oxford University Press.

United Nations. 2007. *The Wye Group Handbook: Rural Households' Livelihood and Well-Being: Statistics on Rural Development and Agriculture Household Income*. Geneva: UN.

UNDP. 1990. *Human Development Report 1990*. New York: UNDP.

Chapter 4
Localizing SDGs as SDGs Desa

Abstract There are at least three strategies to localize development: (1) the civic approach, identifying beneficiaries by name and address; (2) placing development in areas that can be micro-managed, such as villages; and (3) using a multi-stakeholder participatory strategy. Interestingly, this strategy is consistent with the Sustainable Development Goals (SDGs), so it is possible to localize the grand narrative of the SDGs at the village level.

> The village is our future
> This belief comes to me
> Because I don't want calamity
> *Iwan Fals, lyrics of* Village

The concept of development has always been in the form of a grand narrative on a large scale, so it becomes a challenge to localize it, especially at the village level. In its initial formulation (Esteva 1992), development is viewed as a positive change, especially in transforming developing countries into developed countries. When development is placed at the national level, it is measured by indicators such as Gross National Product (GNP), clearly presenting a measure of national economic development in each country.

The grand narrative was narrowed in scope when the focus on national development shifted toward local or regional development (Higgins and Savoie 2005; Sakamoto 2007; Shankar and Shah 2001). The term 'regional' can refer to regional progress between neighboring countries (Togo 2001), such as the Southeast Asian region. However, in this context, 'regional' refers to the various regions that make up a country. In Indonesia, for example, Gross Regional Domestic Product (GRDP) refers to the accumulation of economic-added value obtained from each province. When a measure refers to a province, for example, the GRDP of Papua, it contains the accumulation of economic-added value from all of the regencies therein.

When the concept of development was narrowed to the regional level in Indonesia, it only extended to the regency level. This means data is collected, processed, and analyzed at the regency level. National surveys can only be read at the provincial

A. H. Iskandar, *Accelerating the Achievement of the National Sustainable Development Goals in Indonesia*, Engaging Indonesia, https://doi.org/10.1007/978-981-97-8919-1_4

level, or at least at the regency level (Yusuf 2006). On the other hand, the district GDP, let alone the village GDP, is unknown.

Up to this point, the challenges of village development are clear. Village development is difficult to formulate because of the absence of supporting data at the sub-village level, that is, data at the level of community units (*rukun warga*, RW) and neighborhood units (*rukun tetangga*, RT), as well as civic data from families and individuals. Individual data may exist, such as population identity data, but it has never been formulated into a development concept at the village level. On the other hand, there have been attempts to make proxy village-level poverty data from provincial-level data, but these have not been successful. This is shown when the results of the village-level proxies are accumulated back to the national level, where the results are different from the initial data. What has been done regarding village development is ultimately limited to making references to big concepts and then applying them to the community at the village level.

4.1 Globalization of Development

Development has broadened the meaning of social change, which is no longer defined solely as a result of processes from within the community itself but also involves external influences (Suwarsono and So 2014). External influence can be viewed positively, as a means of modernizing a country that is still lagging behind. Conversely, external influences can also be seen as increasing the dependence of underdeveloped countries on more developed countries.

There are two main schools of thought on international development, and both continue to be influential. The first view, which emerged at the end of the Second World War and strengthened in the 1950s, is that international development is the modernization of underdeveloped countries (Suwarsono and So 2014). Developed countries, especially the victors in the Second World War, became the point of reference for development goals. The United States, which was the most developed and wealthy country at that time, initially provided economic and technical assistance to European countries to rise from the damage and adversity caused by the war. After American support had enabled Europe to rebuild, the US began to follow an international development program, providing assistance more widely to developing countries. There were successful examples, such as South Korea, which was supported by projects that were a key component of US foreign policy. Other developed nations also followed, providing development assistance to undeveloped countries.

From this point of view, developed countries are not only a reference for material aspects but also for moral development. Religious teachings that open opportunities to enter heaven or the highest goodness for every human being who continues to strive are believed to be a source of energy for developing culture. This can be seen in both the Protestant ethic and from the religious values that originated from pre-industrial Japan's Tokugawa period (Bellah 1992; Weber 2004).

Indeed, among the adherents of modernization are those who seek to show the direction of progress, not only by capitalism in Europe and the United States, but also a path of progress led by communism and fascism (Moore 1966). The progress of the Soviet Union (now Russia and other Eastern European countries) referred to the role of people's communes, rather than the market, to streamline agricultural development toward industry. As for the progress of Germany and Japan, they dared to ignite the Second World War, showing fascism is also a path of modernity, wherein absolute leadership accumulates power that obliges citizens to support the will of elite development.

The second school of thought on international development is that rather than leading to modernization, it actually retards the progress of nations, making them dependent on developed countries (Suwarsono and So 2014). For example, Argentina was originally a developed country but was downgraded after incurring debt to donor countries. Unfortunately, while a country is downgraded in status to developing or less-developed, its elite usually maintain good relations with the elite of the donor countries, as compradors representing the interests of developed countries in their own countries.

Modernization and dependency occupy the two sides of opposing development paradigms. In many cases of development, the same facts can be interpreted in opposite directions. If a developing country falls in a development index ranking, adherents of the dependency paradigm will blame external failure, arguing that its surplus has flowed to developed countries. Conversely, modernization adherents will blame internal failure, arguing the country's ranking fell because of its weak capacity.

Both the modernization and dependency paradigms present an analysis of the globalization of development, as interaction takes place between countries at an international level. The role of the state is clearly dominant in national development.

The dichotomy of the modernization versus dependency paradigms has increasingly been circumvented by the globalization of development, which has been strengthened through agreements between nations, under the auspices of the United Nations (Basile and Baud 2019). This is where the grand narrative gets its most fruitful models of development. Various high-level development conferences have been held by the United Nations to drive global social change. Development summits on climate and ecology have been held since the 1970s. The global aspect of ecological development is indeed easier to embrace in line with the accumulation of knowledge, as the ecosystems in each country are interconnected to form the world's biosphere. Forest fires in Sumatra, for example, affect the lives of Singaporeans and Malaysians. The reduced area of tropical rainforests in Brazil and Indonesia contributes to the melting of the polar icecaps.

The United Nations hosted social development and population development conferences in the 1990s, discussing disaster risks to development, and related matters. It is from the accumulation of lessons gathered from the dimensions of global development that the Sustainable Development Goals (SDGs) were formed (UN 2015). Thus, the SDGs can be regarded as the accumulation of human knowledge on development.

The SDGs can also be viewed as a means of governance of global development by integrating international development in line with the goals (Kanie et al. 2017). However, there are weaknesses in the legality of the SDGs: their implementation is voluntary for UN members, and they lack strict governing institutions with various localizations (Biermann and Kanie 2017). Nevertheless, the strength of the SDGs is in the uniformity of measuring the achievement of global development goals. The achievements of each country are then compared and ranked. No country wants to be in last position, so there is motivation to achieve a high ranking while at the same time improving life within the country itself.

4.2 Localizing Development Through People's Participation

One of the criticisms of the modernization and dependency paradigms is the dominant role of the state. Development is seen primarily as being dominated by the elite in each country. Such criticism prompted the formulation of a counter-paradigm that prioritizes people's participation (Korten 2001). Starting in the 1970s, a people-centered development approach emerged.

When compared with the previous two paradigms, which prioritized the state and then supported large institutions and the private sector, even multinational private companies, the new focus of people-centered development targets a third sector of development: society (Cernea, ed. 1985). The agents of this development include non-governmental organizations (NGOs), non-governmental development organizations (known in Indonesia by the abbreviation LPSM, which were more common in the 1970s and 1980s), community organizations, self-help groups, and community groups (Sajogyo 1997; Verhagen 1996).

Initially, people-centered development prioritized building the capacity or education of community groups, rather than prioritizing infrastructure and economic development. The group becomes a place for dynamic interactions between community members, so that the interaction makes it easier for citizens to learn from their own circles (Siregar 2001). Learning from close friends and fellow citizens, in every unique way, accelerates understanding of the changing environment. Moreover, it is believed that the success of development is determined more by the readiness of citizens to know, understand, and practice various joint activities (Uphoff 1985).

This participation paradigm becomes a bridge that emphasizes locality and its translation into development at the village level, even directing development starting from the village, known as 'building from behind' (Chambers 1987). It is important to note the development of this participatory paradigm is not primarily based on the substance of development, except for the substance of community capacity. Instead, it is based on the method of development, namely, the participatory method for development planning (Chambers 1994; Mikkelsen 1999). Aspects of the participation paradigm include Rapid Rural Appraisal, Participatory Rural Appraisal, Participatory Learning and Action, participatory surveys, and so on.

All of these participatory methods are aimed at building consensus among citizens (Coleman 1994). Further research finds the essence of participation lies in citizens' decision-making (Arnstein 1969). Consensus on planning can give birth to participation initiatives in its implementation, so that all citizens benefit from development outcomes together (Uphoff 1985).

The emphasis on people raises the need for data on individuals, families, and small groups to increase development effectiveness (UN 2007). Their real data, by name and address, becomes important. They are the beneficiaries of development, and knowing the beneficiaries from the planning stage can ensure they enjoy the benefits of development outcomes (Uphoff 1985), thereby increasing the effectiveness of development. Territoriality refers to the attachment to a geographic space by certain peoples, such as individuals who are members of a family. In Indonesia, about ten families make up a small community unit called a *dasawisma*. Several *dasawisma* form a neighborhood unit (RT), several RT form a community unit (RW), and several RW form a village. Thus, when a plan is made to build a road in the village, it cannot be assumed that all villagers will receive the benefits; instead, when villagers are known by name and address, the beneficiaries can be more clearly identified.

Unfortunately, the emphasis on this method is difficult to accommodate into knowledge of the substance of local development. For example, it is not easy to formulate that village development must include infrastructure aspects, because the substance is not determined by a development theory but by local consensus decisions. For example, if residents decide that improving their skills in mining is the most important priority, then such training becomes the substance of local village development, rather than projects such as the building of roads.

4.3 Localizing the SDGs in the Village

The dependency paradigm starts from fulfilling infrastructure before building a community structure, culminating in the development of a superstructure such as culture. Physical conditions influence interactions between people, and then social relationships affect ways of thinking (Sanderson 2010).

The modernization paradigm has a different direction, with the starting point emphasizing a rational mindset, an entrepreneurial culture, which is understood as a driver of action. These culturally directed patterns of action among a group of people influence the structural patterns of society and ultimately determine the way in which its territory is developed.

Meanwhile, other groups have other development priorities. Environmental groups are more oriented toward nature conservation, to the point of rejecting development that results in pollution or the extinction of plant and animal species. Disaster mitigation groups direct the substance of development toward disaster response tools, materials, and training. Population groups focus more attention on family planning before developing other substances. The participation paradigm focuses more on the processes in planning, implementing, and obtaining development outcomes.

In order to avoid falling into a tug-of-war between the substances and processes of development, one strategy is to ensure that the entire substances and processes of development exist within a defined area or among a group of people. The development goals can then be directed to fulfill various substances and processes directed by previous development concepts. It is at this point that the SDGs take on a central role. Development goals comprise various aspects that must be met in development (Kanie et al. 2017; UN 2015). This refers to the desired outputs, outcomes, and impact. In particular, there are goals directed toward the benefit of people, without anyone being left behind, namely: no poverty, zero hunger, good health and well-being, quality education, gender equality, and clean water and sanitation. The goals directed at the achievement of regional development include: affordable and clean energy; decent work and economic growth; industry, innovation, and infrastructure; reduced inequality; sustainable cities and communities; responsible production and consumption; climate action; life below water; and life on land. The goals of the expected development process are: peace, justice, and strong institutions; and partnerships to achieve the goals.

From the discussion in this chapter, there are at least three strategies to localize development: (1) the civic approach, identifying beneficiaries by name and address; (2) placing development in areas that can be micro-managed, such as villages; and (3) using a multi-stakeholder participatory strategy. Interestingly, this strategy is consistent with the SDGs, as explained in Table 4.1, so it is possible to localize the grand narrative of the SDGs at the village level.

The opportunity for localization of the SDGs has actually existed since the publication of the goals (UN 2015). The aim of localization is to garner greater participation and contributions from more parties, thereby accelerating the achievement of the 2030 agenda for development. In fact, there are signs that localization is a prerequisite for the successful implementation of the SDGs (Servaes 2017). In Indonesia, the global SDGs were localized into the National SDGs (known as *Tujuan Pembangunan Berkelanjutan*, TPB) under Presidential Decree 59/2017. The National SDGs were then further localized at the provincial and regency levels. In addition, localization is also directed at certain groups, such as SDGs for the private sector (Abdulai et al. 2020).

The following five key points can be learned from localization efforts:

Table 4.1 Aspects of the SDGs and localization development strategy

No	Localization development strategy	SDGs
1	Clear beneficiaries in the field, by name and address, as the civic aspect	Numbers 1, 2, 3, 4, 5, 6
2	Locations with an area that can be managed, as the regional aspect	Numbers 7, 8, 9, 10, 11, 12, 13, 14, 15
3	Broad participatory strategy	Numbers 16, 17

1. Localization is formalized in government policy, both at the national and regional levels.
2. Localization involves formulating the goals from English to Indonesian. This is to bring the SDGs closer to the cultural development of the local community.
3. Additional indicators of achievement can be made in the form of local indicators. Basically, indicators are prepared so that comparisons can be made with other countries, thus forming a global development achievement.
4. For certain groups, such as the private sector, localization can include the formulation of specific indicators that stimulate their participation in development. For example, private partnerships in development, with an indicator being infrastructure development that yields long-term results. So far, the private sector's role is more in the form of corporate social responsibility contributions (Crowther et al. 2018).
5. There have been numerous proposals to add an 18th goal to the SDGs.[1] This has been done both informally and seriously (Visseren-Hamakers 2020), with an 18th goal even entering government regulations, such as Indonesia's SDGs Desa.

For Indonesia's villages, the localization of the SDGs into the SDGs Desa was essential. In fact, the SDGs Desa are the main reference for the medium-term development of villages throughout the country. The proven SDGs make it easier to measure development. Each measure comprehensively covers aspects of people's lives and their environment. Therefore, localizing the SDGs as the SDGs Desa makes the direction of village development clear and detailed through holistic goals.

The creation of the SDGs Desa means the SDGs have been Indonesianized and localized at the village level. The use of language in the SDGs Desa is important: it must be simple, refer to a statement of purpose, and be rational to fulfill, if development is to be carried out. Table 4.2 shows the wording that was chosen to emphasize the objectives to be achieved by the SDGs Desa through development. For example, instead of 'no poverty', or 'eliminating all forms of poverty', the appropriate wording is: 'villages without poverty'. This wording is easier for villagers to accept because it directly refers to village-level development. It also refers to a statement of purpose, beginning with a noun, rather than a process starting with a verb (in Indonesian, the prefix '*me-*') or a quantifier.

Although the indicators will be described in detail in the next chapter, it can be stated here that the indicators were derived to be measured only at the village level. This constituted a major leap, yet there is no other way if village development is to be realized systematically. Separate measurements must be made at the village level, so that the data collected can be processed for village-level development, as well as being accumulated into regional and national indicators.

When the indicators for the SDGs Desa were being formulated, it was known that one of the shortcomings of the SDGs was the absence of development indicators in the context of conditions typical of Indonesian villages (Koentjaraningrat 1971; Laksono

[1] Proposals for an 18th SDG have ranged from 'love and joy' to nuclear disarmament, development of the space economy, and animal rights.

Table 4.2 Localizing the SDGs' target statements into the SDGs desa

SDGs	National SDGs	SDGs Desa
1. End poverty in all its forms everywhere	1. End poverty in all its forms everywhere	1. Village without poverty
2. End hunger, achieve food security and improved nutrition and promote sustainable agriculture	2. Eliminate hunger, achieve food security and good nutrition, and promote sustainable agriculture	2. Village without hunger
3. Ensure healthy lives and promote well-being for all at all ages	3. Ensure healthy lives and improve the welfare of the entire population of all ages	3. Healthy and prosperous village
4. Ensure inclusive and equitable quality education and promote lifelong learning opportunities for all	4. Ensure inclusive and equitable quality education and enhance lifelong learning opportunities for all	4. Quality village education
5. Achieve gender equality and empower all women and girls	5. Achieve gender equality and empower females	5. Involvement of village women
6. Ensure availability and sustainable management of water and sanitation for all	6. Ensure availability and sustainable management of clean water and sanitation for all	6. Village with Clean Water and Sanitation
7. Ensure access to affordable, reliable, sustainable, and modern energy for all	7. Ensure access to affordable, reliable, sustainable, and modern energy for all	7. Clean and renewable energy village
8. Promote sustained, inclusive, and sustainable economic growth, full and productive employment, and decent work for all	8. Promote inclusive and sustainable economic growth, productive and full employment opportunities, and decent work for all	8. Equitable village economic growth
9. Build resilient infrastructure, promote inclusive and sustainable industrialization, and foster innovation	9. Build resilient infrastructure, promote inclusive and sustainable industries, and foster innovation	9. Needs-based village infrastructure and innovation
10. Reduce inequality within and among countries	10. Reducing inequality within and among countries	10. Village without gaps
11. Make cities and human settlements inclusive, safe, resilient, and sustainable	11. Make cities and settlements inclusive, safe, resilient, and sustainable	11. Safe and comfortable village residential areas
12. Ensure sustainable consumption and production patterns	12. Ensure sustainable production and consumption patterns	12. Environmentally conscious village consumption and production

(continued)

Table 4.2 (continued)

SDGs	National SDGs	SDGs Desa
13. Take urgent action to combat climate change and its impacts	13. Take urgent action to address climate change and its impacts	13. Climate change responsive village
14. Conserve and sustainably use the oceans, seas, and marine resources for sustainable development	14. Conserve and sustainably use marine and ocean resources for sustainable development	14. Village caring for the marine environment
15. Protect, restore, and promote sustainable use of terrestrial ecosystems, sustainably manage forests, combat desertification, and halt and reverse land degradation and halt biodiversity loss	15. Protect, restore, and promote sustainable use of terrestrial ecosystems, sustainably manage forests, stop desertification, reverse land degradation, and halt biodiversity loss	15. Village caring for the land environment
16. Promote peaceful and inclusive societies for sustainable development, provide access to justice for all and build effective, accountable, and inclusive institutions at all levels	16. Strengthen inclusive and peaceful societies for sustainable development, provide access to justice for all, and build effective, accountable, and inclusive institutions at all levels	16. Village of peace and justice
17. Strengthen the means of implementation and revitalize the global partnership for sustainable development	17. Strengthen the means of implementation and revitalize the global partnership for sustainable development	17. Partnerships for village development
-	-	18. Dynamic village institutions and adaptive village culture

2009; Rubianto 1996). The purpose of placing the SDGs concept at the Indonesian village context is to ensure that the pace of village development does not disappear due to global, national, and regional measurements (Li 2002; Sajogyo 1977). For example, the distinctive Indonesian village features of the vital role of religion and social piety need to be emphasized (Duc 2017). Likewise, the performance of typical institutions that have emerged in villages in Indonesia, such as Village-Owned Enterprises. From this need, SDGs Desa 18 was formulated: Dynamic Village Institutions and Adaptive Village Culture.

In order to facilitate communication in the village, the designated term for all these village development goals is the SDGs Desa. The term 'SDGs' was kept because it is concise and more memorable, even among the villagers themselves, rather than a long and atypical wording, such as 'the Village-Level Sustainable Development Goals'.

The initial dissemination of the SDGs Desa indicated high enthusiasm to understand them, especially through webinars. This shows the importance of choosing the wording of the SDGs Desa. The additional word 'Village' refers to village development, which can be formulated independently in each village, through specific SDGs Desa data, not borrowed data from outside the village. The availability of individual, family, and neighborhood unit data in the village allows the formulation of internal village revenue, which contains added value for economic activities in the village. This is possible because there are many economic business entities in the village, especially those compiled from the businesses of tens to thousands of families in one village. Of course, the detailed formulation of village development will be wide open. All of this is part of the effort to develop a smaller, more local, village-level development concept, directly to the residents, with a participatory main approach, which is practical to use, so that villagers can quickly benefit from it.

Just as participatory methods attach great importance to images because they can show more clearly the knowledge and needs of villagers, the SDGs icons are also localized in the SDGs Desa (Table 4.3). For this reason, the images have been rearranged to be clearer, less abstract, and closer to the reality in Indonesian villages. For example, SDG 2, Zero Hunger, was Indonesianized as 'an end to hunger', at the national level, while in the SDGs Desa it was phrased as 'Village Without Hunger'. Instead of using the icon of a steaming bowl, the SDGs Desa chose a rice basket icon, which is more common in villages in Indonesia. The Indonesian icons facilitate the campaign of the SDGs Desa campaign in 75,265 villages in the archipelago.

The localization of the SDGs into the SDGs Desa was formulated as follows:

1. SDGs Desa 1: Village Without Poverty. This wording was chosen over No Poverty or 'ending poverty'. The chicken-shaped moneybox icon shows that villagers have the trait to combat poverty by starting a savings habit, even if their savings are kept at home.
2. SDGs Desa 2: Village Without Hunger. This wording was chosen over Zero Hunger or 'ending hunger'. The icon of a woven basket for cooked rice signifies the release of residents from hunger because they can eat rice, which is the largest percentage of village household expenditure.
3. SDGs Desa 3: Healthy and Prosperous Village. This wording was chosen over Good Health and Well-being. The icon of a heartbeat in a pair of hands signifies that life is supported more humanely.
4. SDGs Desa 4: Quality Village Education. This wording was chosen over Quality Education. The icon of a villager reading, while wearing a traditional conical hat, symbolizes the lifelong education of villagers through continual reading of the latest information.
5. SDGs Desa 5: Involvement of Village Women. This wording was chosen over Gender Equality to highlight the importance of increasing the involvement of women to achieve gender equality in the villages. The icon of gender symbols is similar to that of the SDGs.

Table 4.3 Localizing the SDGs' icons into the SDGs desa

SDGs	National SDGs	SDGs Desa
1 NO POVERTY	1 MENGHAPUS KEMISKINAN	1 VILLAGE WITHOUT POVERTY
2 ZERO HUNGER	2 MENGAKHIRI KELAPARAN	2 VILLAGE WITHOUT HUNGER
3 GOOD HEALTH AND WELL-BEING	3 KESEHATAN YANG BAIK DAN KESEJAHTERAAN	3 HEALTHY AND PROSPEROUS VILLAGE
4 QUALITY EDUCATION	4 PENDIDIKAN BERMUTU	4 QUALITY VILLAGE EDUCATION

(continued)

Table 4.3 (continued)

SDGs	National SDGs	SDGs Desa
5 GENDER EQUALITY	5 KESETARAAN GENDER	5 INVOLVEMENT OF VILLAGE WOMEN
6 CLEAN WATER AND SANITATION	6 AKSES AIR BERSIH DAN SANITASI	6 VILLAGE WITH CLEAN WATER AND SANITATION
7 AFFORDABLE AND CLEAN ENERGY	7 ENERGI BERSIH DAN TERJANGKAU	7 CLEAN AND RENEWABLE ENERGY VILLAGE
8 DECENT WORK AND ECONOMIC GROWTH	8 PEKERJAAN LAYAK DAN PERTUMBUHAN EKONOM	8 EQUITABLE VILLAG ECONOMIC GROWT

(continued)

Table 4.3 (continued)

SDGs	National SDGs	SDGs Desa
9 INDUSTRY, INNOVATION AND INFRASTRUCTURE	9 INFRASTRUKTUR, INDUSTRI DAN INOVASI	9 NEEDS-BASED VILLAGE INFRASTRUCTURE AND INNOVATION
10 REDUCED INEQUALITIES	10 MENGURANGI KETIMPANGAN	10 VILLAGE WITHOUT GAPS
11 SUSTAINABLE CITIES AND COMMUNITIES	11 KOTA DAN KOMUNITAS YANG BERKELANJUTAN	11 SAFE AND COMFORTABLE VILLAGE RESIDENTIAL AREAS
12 RESPONSIBLE CONSUMPTION AND PRODUCTION	12 KONSUMSI DAN PRODUKSI YANG BERTANGGUNGJAWAB	12 ENVIRONMENTALLY CONSCIOUS VILLAGE CONSUMPTION AND PRODUCTION

(continued)

Table 4.3 (continued)

SDGs	National SDGs	SDGs Desa
13 CLIMATE ACTION	13 PENANGANAN PERUBAHAN IKLIM	13 CLIMATE CHANGE RESPONSIVE VILLAGE
14 LIFE BELOW WATER	14 MENJAGA EKOSISTEM LAUT	14 VILLAGE CARING FOR THE MARINE ENVIRONMENT
15 LIFE ON LAND	15 MENJAGA EKOSISTEM DARAT	15 VILLAGE CARING FOR THE LAND ENVIRONMENT
16 PEACE, JUSTICE AND STRONG INSTITUTIONS	16 PERDAMAIAN, KEADILAN, DAN KELEMBAGAAN YANG KUAT	16 VILLAGE OF PEACE AND JUSTICE

(continued)

Table 4.3 (continued)

SDGs	National SDGs	SDGs Desa

-	-	

6. SDGs Desa 6: Village with Clean Water and Sanitation. This wording was chosen over Clean Water and Sanitation. The icon of a traditional water jug is well known by villagers and signifies access to potable drinking water.
7. SDGs Desa 7: Clean and Renewable Energy Village. This wording was chosen over Affordable and Clean Energy. The icon depicts a lightbulb with a leaf, powered by wind turbines. This signifies that villages can have clean and renewable energy by generating electricity from windmills.
8. SDGs Desa 8: Equitable Village Economic Growth. This wording was chosen over Decent Work and Economic Growth. The icon depicts two villagers and a graph of economic growth, showing that growth is part of the village environment.
9. SDGs Desa 9: Needs-Based Village Infrastructure and Innovation. This wording was chosen over Industry, Innovation, and Infrastructure. The icon of a long and winding road reflects the unique needs of the Indonesian archipelago, where village funds have so far been used to develop 121,000 kilometers of roads.
10. SDGs Desa 10: Village Without Gaps. This wording was chosen over Reduced Inequalities. The icon of two villagers on parallel scales indicates equal welfare between residents.
11. SDGs Desa 11: Safe and Comfortable Village Residential Areas. This wording was chosen over Sustainable Cities and Communities. The icon of a typical village house in a moving circle symbolizes the sustainability of the village residential area.
12. SDGs Desa 12: Environmentally Conscious Village Consumption and Production. This wording was chosen over Responsible Consumption and Production.

The icon features a villager placing garbage into a selected trash can, illustrating the importance of separating rubbish so that waste banks can recycle it into productive goods.

13. SDGs Desa 13: Climate Change Responsive Village. This wording was chosen over Climate Action or 'handling climate change'. As climate involves regional and global conditions, the main role of the village is not primarily to handle the issue but to be responsive through climate change mitigation. The icon of children playing in the rain shows that when climate change is managed, the climate is beneficial to the people.
14. SDGs Desa 14: Village Caring for the Marine Environment. This wording was chosen over Life Below Water or 'maintaining marine ecosystems'. The icon of a village fisherman casting his net over a shallow bay signifies that a protected ocean ecosystem makes it easy to catch fish.
15. SDGs Desa 15: Village Caring for the Land Environment. This wording was chosen over Life on Land or 'maintaining terrestrial ecosystems'. The icon of a fertile rice field signifies the productivity of a sustainably managed environment.
16. SDGs Desa 16: Village of Peace and Justice. This wording was chosen over Peace, Justice and Strong Institutions. The *kentongan* (wooden slit drum) icon symbolizes a safe, peaceful, and just village.
17. SDGs Desa 17: Partnerships for Village Development. This wording was chosen over Partnerships for the Goals. The icon represents how the relationship between residents and other parties can create beneficial partnerships.
18. SDGs Desa 18: Dynamic Village Institutions and Adaptive Village Culture. This additional goal signifies the uniqueness of Indonesian village development. The icon symbolizes village development in line with the priorities set by the Ministry of Villages, Development of Disadvantaged Regions, and Transmigration.

References

Abdulai, D., O. Knauf, and L. O'Riordan. 2020. Achieving Sustainable Development Goals 2030 in Africa: A Critical Review of the Sustainability of Western Approaches In *The Future of the UN Sustainable Development Goals: Business Perspectives for Global Development in 2030*, ed. S.O. Idowu, R. Schmidpeter, and L. Zu. Switzerland: Springer.

Arnstein, S. R. 1969. A Ladder of Citizen Participation. *Journal of the American Institute of Planners* 35: 4.

Basile, E., and I. Baud. 2019. Introduction: Reflecting on Development Studies in the New Millennium. In *Building Development Studies for the New Millennium*, ed. I. Baud, E. Basile, T. Kontinen, and S. V. Itter. Cham, Switzerland: Palgrave Macmillan.

Bellah, R. N. 1992. *Religi Tokugawa: Akar-akar Budaya Jepang*. Indonesian translation. Jakarta: Gramedia.

Biermann, F., and N. Kanie. 2017. Conclusion: Key Challenges for Global Governance through Goals. In *Governing through Goals: Sustainable Development Goals as Governance Innovation*, ed. N. Kanie, and F. Biermann. Massachusetts: MIT Press.

Cernea, M., ed. 1985. *Putting People First: Sociological Variables in Rural Development*. New York: Oxford University Press.

Chambers, R. 1987. *Pembangunan Desa: Mulai dari Belakang*. Indonesian translation. Jakarta: LP3ES.

Chambers, R. 1994. *PRA (Participatory Rural Appraisal), Memahami Desa Secara Patisipatif*. Indonesian translation from Participatory rural appraisal (PRA): Challenges, Potentials and Paradigm. Yogyakarta: Kanisius.

Coleman, J. S. 1994. *Foundations of Social Theory*. Cambridge: Harvard University Press.

Crowther, D., S. Seifi, and A. Moyeen. 2018. Responsibility and Governance in Achieving Sustainability. In *The Goals of Sustainable Development: Responsibility and Governance*, ed. D. Crowther, S. Seifi, and A. Moyeen. Switzerland: Springer.

Duc, A. L. 2017. Religious Environmentalism and Environmental Sustainability in Asia. In *Sustainable Development Goals in the Asian Context*, ed. J. Servaes. Singapore: Springer.

Esteva, G. 1992. Development. In *The Development Dictionary: A Guide to Knowledge as Power*, ed. W. Sachs. London: Zed Books.

Higgins, B., and D. J. Savoie. 2005. *Regional Development Theories and Their Application*. New Brunswick: Transaction.

Kanie, N., S. Bernstein, F. Biermann, and P. M. Haas. 2017. Introduction: Global Governance through Goal Setting. In *Governing through Goals: Sustainable Development Goals as Governance Innovation*, ed. N. Kanie and F. Biermann. Massachusetts: MIT Press.

Koentjaraningrat, R. M. 1971. *Manusia dan Kebudayaan di Indonesia*. Jakarta: Djambatan.

Korten, D. C. 2001. *Menuju Abad Ke-21: Tindakan Sukarela dan Agenda Global*. Indonesian translation. Jakarta: Yayasan Obor Indonesia.

Laksono, P. M. 2009. *Tradisi dalam Struktur Masyarakat Jawa Kerajaan dan Pedesaan, Alih Ubah Model Berpikir Jawa*. Yogyakarta: Kepel Press.

Li, T. M. 2002. Keterpinggiran, Kekuasaan dan Produksi: Analisis terhadap Transformasi Daerah Pedalaman. In *Proses Transformasi Daerah Pedalaman di Indonesia*, ed. T.M. Li. Jakarta: Yayasan Obor Indonesia.

Mikkelsen, B. 1999. *Metode Penelitian Partisipatoris dan Upaya-Upaya Pemberdayaan*. Indonesian translation from Methods for Development Work and Research: A Guide for Practitioners. Jakarta: Yayasan Obor Indonesia.

Moore, B. J. 1966. *Social Origins of Dictatorship and Democracy: Lord and Peasant in the Making of The Modern World*. Boston: Bacon Press.

Rubianto, V. 1996. *Paradigma Asia: Pertautan Kemiskinan dan Kereligiusan dalam Teologi Aloysius Pieris*. Yogyakarta: Kanisius.

Sajogyo. 1977a. Golongan Miskin dan Partisipasinya dalam Pembangunan Desa. In *Prisma*, vol. 6, No. 3.

Sajogyo. 1997b. *Menciptakan Visi Mendukung Pengembangan Kelompok Swadaya Mandiri dalam Gerakan Nasional PPK*. Jakarta: Puspa Swara.

Sakamoto, H. 2007. *The Dynamics of Inter-Provincial Income Distribution in Indonesia*. Working Paper Series Vol. 2007–25. Kitakyushu: The International Centre for the Study of East Asian Development.

Sanderson, 2010 Sanderson, S. K. 2010. *Sosiologi Makro*. Indonesian translation. Jakarta: Rajawali Press.

Shankar, R., and A. Syah. 2001. Bridging the Economic Divide within Nations: A Scorecard on the Performance of Regional Development Policies in Reducing Regional Income Disparities. World Bank Policy Research Working Paper No. 2717.

Siregar, B. B. 2001. *Menelusuri Jejak Ketertinggalan, Merajut Kerukunan Melintasi Krisis*. Bogor: Pusat P3R-YAE.

Suwarsono, A., and Y. So. 2014. *Perubahan Sosial dan Pembangunan*. Jakarta: LP3ES.

Togo, K. 2001. *A Brief Survey on Regional Convergence in East Asian Economies*. Musashi University Working Paper No. 5.

United Nations. 2007. *The Wye Group Handbook: Rural Households' Livelihood and Well-Being: Statistics on Rural Development and Agriculture Household Income*. Geneva: UN.

United Nations. 2015. *Transforming Our World: The 2030 Agenda For Sustainable Development*. New York: UN.

Uphoff, N. 1985. Fitting Projects to People. In *Putting People First: Sociological Variables in Rural Development*, ed. M. Cernea. New York: Oxford University Press.

Verhagen, Koenraad. 1996. *Pengembangan Keswadayaan: Pengalaman LSM di Tiga Negera*. Indonesian translation. Jakarta: Penebar Swadaya.

Visseren-Hamakers, I. J. 2020. *The 18th Sustainable Development Goal*. In *Earth System Governance*, No. 3.

Weber, M. 2004. *Etika Protestan dan Spirit Kapitalisme*. Indonesian translation. Yogyakarta: Pustaka Pelajar.

Yusuf, A. A. 2006. *On the Re-assessment of Inequality in Indonesia: Household Survey or National Account?* Paper submitted to the 8th IRSA International Conference, 18–19 August, 2006, Malang, Indonesia.

Chapter 5
SDGs Desa

Abstract As a signatory to the Sustainable Development Goals (SDGs), Indonesia is taking strategic steps, policies, and various activities to achieve the SDGs by 2030. A new policy was needed that could reformulate Presidential Regulation No. 59 of 2017, which regulates the implementation of the Sustainable Development Goals (SDGs) to the provincial and regency/municipal levels, so the goals could be channeled down to the village level named SDGs Desa.

As a signatory to the Sustainable Development Goals (SDGs), Indonesia is taking strategic steps, policies, and various activities to achieve the SDGs by 2030. Presidential Regulation No. 59 of 2017 on the Implementation of Achieving the Sustainable Development Goals serves as the legal umbrella. It also provides direction for the Indonesian people, starting from the central government, regional governments, as well as funders and the private sector, to work together to achieve the SDGs.

The SDGs are a universal set of goals, targets, and indicators for sustainable development. For this reason, the indicators of the SDGs can be applied to measure the reality of development at the global, regional, national, local, and even village levels.

Explicitly, the SDGs are intended to mobilize development that results in progress, especially through efforts to end all forms of poverty and hunger everywhere, reduce inequality within and between countries, preserve the environment, and take strategic steps on climate change.

These goals are represented in the four pillars of the SDGs, which are interrelated with each other. The first pillar is the social pillar. Through this pillar, the SDGs are intended to achieve development oriented toward the fulfillment of basic human rights in a quality, fair, and equal manner, as well as improving the welfare of the entire community.

If I see village children's faces with shining eyes, 'Pak Merdeka, Pak Merdeka, Pak Merdeka!', (Freedom, Sir, Freedom, Sir) I no longer see human eyes, I see Indonesia.

Soekarno

A. H. Iskandar, *Accelerating the Achievement of the National Sustainable Development Goals in Indonesia*, Engaging Indonesia, https://doi.org/10.1007/978-981-97-8919-1_5

The second pillar is the economy. To achieve the social pillar, the economic pillar is needed, providing a way for development to achieve quality economic growth through sustainable employment and business opportunities, innovation, inclusive industrialization, adequate infrastructure, affordable clean energy, and the support of partnerships.

In turn, the economic pillar cannot be separated from the third pillar, namely the environmental pillar. Economic growth is created to fulfill basic human rights, achieved through sustainable management of natural resources and the environment, in order to meet the needs of today's society and to ensure the needs of future generations.

In view of this, the fourth pillar of the SDGs is law and governance, which can provide legal certainty and the achievement of effective, transparent, accountable, and participatory governance for the creation of security stability as a prerequisite for effective and sustainable development.

5.1 SDGs Desa Action

As a joint action for the planet, people, and prosperity, as well as to strengthen universal peace and freedom, the SDGs require the participation of the global population to jointly foster the acceleration of the achievement of the goals.

Presidential Regulation No. 59 of 2017 on the Implementation of Achieving the Sustainable Development Goals is a manifestation of the Indonesian government's commitment to be part of the global community's joint action in promoting sustainable development to overcome poverty, hunger, inequality, education gaps, and climate change. This Presidential Regulation is the guideline for the convergence of the SDGs with national development plans.

The convergence of the SDGs in national development planning policies is a follow-up to the UN General Assembly's resolution titled *Transforming Our World: The 2030 Agenda for Sustainable Development*. In addition to agreeing on the 17 SDGs, the assembly also agreed to the integration of global development goals into the national development policies of each member country.

In Indonesia, the SDGs convergence policy began with ratifying the SDGs through Presidential Regulation No. 59 of 2017. This Presidential Regulation mandated the preparation of a National Roadmap for the Sustainable Development Goals, and a National Action Plan for the Sustainable Development Goals. The regulation also mandated the preparation of a Regional Action Plan for the Sustainable Development Goals.

The regulation shows that, first, the Indonesian government realized it would be impossible to achieve the SDGs up to 2030 if supporting policies were only at the national level. This means that joint action on the SDGs must involve many elements, including local governments, both provincial and regency/municipal. Second, the government acknowledged the regions possess wisdom that has accumulated over time. In addition, the nation's regions have a diversity of cultures and socio-economic conditions, as well as diversity in infrastructure and facilities, resulting in the need for the Regional Action Plan for the Sustainable Development Goals.

Villages, as regulated in Law No. 6 of 2014, are formulated as legal community units that have territorial boundaries and are authorized to regulate and manage their government affairs and the interests of the local community based on community initiatives, origin rights, and/or traditional rights that are recognized and respected in the system of government of the Unitary State of the Republic of Indonesia.

Implementation of Law No. 6 of 2014 gives villages wide opportunities to support the acceleration of achieving the SDGs. In addition to having the authority to manage their own households and access village funds, when viewed from the regional and civic aspects, the village has the potential to be a determinant of the achievement of the SDGs by 2030 in Indonesia. Based on these aspects, villages will contribute 74% of the achievement of the SDGs.

Therefore, a policy of centralizing the planning and implementation of village development was needed to accelerate the achievement of the SDGs. This means the SDGs stated in Presidential Regulation No. 59 of 2017 required elaboration at the village level and were then integrated into village development planning, thus forming the SDGs Desa.

For this reason, a new policy was needed that could reformulate Presidential Regulation No. 59 of 2017, which regulates the implementation of the SDGs to the provincial and regency/municipal levels, so the goals could be channeled down to the village level.

5.2 SDGs Desa 18: Dynamic Village Institutions and Adaptive Village Culture

Recent data on development activities and results shows that rural village areas have been lagging behind urban areas. This lag is mostly due to the state's low level of attention to villages. In fact, the village holds a strategic position in solving various development problems. For this reason, an ambitious and transformational vision of village development needed to be established.

Referring to the National SDGs, village development targets until 2030 can be set so that villages can achieve: freedom from poverty, freedom from hunger, healthy people, freedom from fear and violence, universal literacy, safe drinking water and sanitation, securely resilient and sustainable human habitats, universal respect for the human rights and dignity of all residents, rule of law, justice, equality, non-discrimination, and respect for race, ethnicity, and cultural diversity.

Thus, the SDGs Desa are an integrated effort to accelerate the achievement of the National SDGs through strengthening various typologies of villages, namely: villages without poverty and hunger, economic villages that grow equitably, villages that care about health, villages that care about the environment, villages that care about education, women-friendly villages, networked villages, and culturally responsive villages.

When these eight village typologies can be achieved in one village at the same time, that's when the village can be called a 'Pancasila Village'. A Pancasila Village is the embodiment of a village that is able to achieve the SDGs Desa simultaneously, being free from poverty and hunger, having partnerships that can support village development activities, having quality education for residents, being healthy and prosperous, having good governance, and being responsive to culture and global climate change.

Thus, the ultimate goal of village development toward 2030, when the SDGs should be achieved, is the realization of the Pancasila Village. Consequently, all the resources, potential, efforts, and endeavors of this nation are directed as much as possible toward realizing this desired village. As a constituent of the state, the village really does determine the face of Indonesia. Therefore, when villages without poverty, without hunger, and so on are realized, Indonesia will be free from hunger and poverty in 2030. This means that achieving the SDGs Desa is the key to achieving the SDGs.

5.3 Dimensions of the SDGs Desa

As a derivative of the global SDGs and the National SDGs, both built on five dimensions, the SDGs Desa are also built on the same dimensions, with an additional dimension, making the following six dimensions:

1. Prosperous Villagers

   The SDGs Desa are to ensure that all villagers can be free from poverty and hunger in all dimensions, and to ensure that all villagers can fulfill their potential with dignity by having equality and the right to live with dignity in a healthy environment.
2. Balance of Nature

   The SDGs Desa seek to protect nature from the adverse impacts of human activities, such as environmental change and irresponsible use of natural resources, for the sake of supporting the needs of present and future generations.
3. Village Prosperity

   The SDGs Desa are to ensure that all villagers can have a prosperous, self-sufficient life and that economic, social, and technological progress can exist in harmony with nature.
4. Peace

   The SDGs Desa seek to foster the growth of peaceful, just, and inclusive villagers who are free from fear and violence. There can be no sustainable development without peace and social security, and vice versa, there can be no peace and social security without sustainable development.
5. Partnerships

The SDGs Desa are to mobilize the necessary means for villages to collaborate together and/or cooperate with third parties for the success of sustainable development, which can only be achieved through close cooperation based on the principle of strong solidarity to achieve competitive economic values. Partnerships with third parties are carried out by villages to accelerate and improve village governance, implementation of village development, village community development, and village community empowerment.

6. Institutional and Village Wisdom

   The SDGs Desa exist to ensure the strength of village institutions, as well as the recognition, respect, and protection of the rights of indigenous peoples, in order to realize strategies to empower traditional customs, local wisdom, and village culture. The SDGs Desa are present as an effort or process of change to achieve a higher level of welfare or quality of life among village communities with high social piety, as well as respect for local customs, culture, and local community beliefs.

These six dimensions become the colors of the SDGs Desa. In addition, the SDGs Desa are linked or have an interdependent relationship with one another, so the achievement of one goal will have an effect on other goals.

The objectives of the SDGs Desa are as follows:

1. Village without poverty.
2. Village without hunger.
3. Village caring for health.
4. Quality village education.
5. Involvement of village women.
6. Village with clean water and sanitation.
7. Clean and renewable energy village.
8. Equitable village economic growth.
9. Needs-based village infrastructure and innovation.
10. Village without gaps.
11. Safe and comfortable village residential areas.
12. Environmentally conscious village consumption and production.
13. Climate change responsive village.
14. Village caring for the marine environment.
15. Village caring for the land environment.
16. Village of peace and justice.
17. Partnerships for village development.
18. Dynamic village institutions and adaptive village culture (Fig. 5.1).

The 18 SDGs Desa are described below.

Fig. 5.1 SDGs desa

5.4 SDGs Desa 1: Village Without Poverty

This goal targets rural village poverty of 0% by 2030. This means there should be no poor people in the village by that time. To make this target a reality, many policies must be taken and implemented by the central, regional, and village governments, such as increasing the income of the poor, ensuring access to basic services, and protecting the entire community from all forms of disaster.

Currently, poverty is still a problem for almost all countries in the world. Therefore, the main agenda of the SDGs is to eliminate global poverty in all its forms and dimensions by 2030.

In the government's Long-Term National Development Plan for 2005–2025, the problem of poverty is seen in a multidimensional framework; therefore, poverty is not only related to income level, but also involves other matters, including: (i) the susceptibility and vulnerability of people or communities to become poor and (ii) the presence or absence of the fulfillment of the basic rights of citizens, including whether there are differences in the treatment of individuals or groups of people regarding their right to live with dignity.

To measure poverty, Statistics Indonesia uses the metric of the ability to meet basic needs. With this approach, poverty is seen as the economic inability to meet basic food and non-food needs, which are measured from the expenditure side. So, the poor are people whose average monthly per capita expenditure is below the poverty line.

The number of poor people in Indonesia in March 2020 reached 26.42 million people. This was an increase of 1.28 million people from March 2019. The percentage of poor people in March 2020 was 9.78%, an increase of 0.37% from March 2019. Meanwhile, based on the area of residence, the percentage of the poverty rate in

Table 5.1 Poverty by area, March 2019–March 2024. *Source* Statistics Indonesia 2024a, b

Area/Year	Number of poor people (million)	Poverty rate
Urban		
Maret 2019	9,99	6,69
Maret 2020	11,16	7,38
Maret 2021	12,18	7,89
Maret 2022	11,82	7,5
Maret 2023	11,74	7,29
Maret 2024	11,64	7,09
Rural		
Maret 2019	15,15	12,85
Maret 2020	15,26	12,82
Maret 2021	15,37	13,1
Maret 2022	14,34	12,29
Maret 2023	14,16	12,22
Maret 2024	13,58	11,79
Totals		
Maret 2019	25,14	9,41
Maret 2020	26,42	9,78
Maret 2021	27,55	10,14
Maret 2022	26,16	9,54
Maret 2023	25,9	9,34
Maret 2024	25,22	9,03

rural villages in March 2020 was 12.82%, down 0.03% from March 2019, while the poverty rate in urban areas rose 0.69%, from 6.69% to 7.38% (Table 5.1).

Disaggregation by area of residence shows that urban residents have a higher average monthly per capita expenditure than residents in rural village areas. Not surprisingly, urban dwellers are spending a lower proportion of their income on food, compared to rural villagers, whose monthly expenditure is mostly used to purchase food. In rural village areas, each resident on average spends Rp1,150,674 a month to meet their needs, namely food (Rp 634,070) and non-food (Rp 531,051). Conversely, urban residents spend more than Rp1 million every month (Rp 1,674,830/capita/month). Of this amount, only 47% is used to buy food commodities (Table 5.2).

Table 5.2 Food expenditure by area of residence, September 2023

Area type	Expenditure on food (%)
Urban and Rural	48,99
Rural	56.38
Urban	45.47

Source Statistics Indonesia 2023a, b, c

In order to achieve the goal of 0% poverty in the village by 2030, there are several targets that must be reached. Among these are: 100% of village communities must have health insurance cards, and all poor people with disabilities and all women-headed households must receive assistance to meet their basic needs. Also, the lowest-income 40% of the village population must receive services necessary for health, including childbirth assistance and immunization, contraception, access to drinking water, and good sanitation. Other targets are access to educational services, decent housing for low-income residents, and the fulfillment of other basic needs.

5.5 SDGs Desa 2: Village Without Hunger

This goal targets the eradication of hunger from villages by 2030, as well as achieving food sovereignty, improved nutrition, and promotion of sustainable agriculture. This goal is in line with Indonesia's development priorities on food security and job creation.

Hunger is often defined as when a person's caloric intake is below minimum energy requirements. In other words, when people lack the food and nutrition security required to lead healthy and productive lives (Fan and Polman, 2014). According to FAO (2008), hunger is a painful sensation caused by food deprivation.

The Global Hunger Index (2019) shows the level of hunger and malnutrition in the world has continued to decline since 2000. This decrease is related to the reduction in the world poverty rate from 28.6% in 1999 to 9.9% in 2015, then to 9.3% in 2017. In the 2023 Global Hunger Index, Indonesia is ranked 77th out of 125 countries. With a score of 17,6 on the Hunger Index, Indonesia is in the 'moderate' category, two points below the 'serious' category.

Average daily calorie consumption per capita in Indonesia in September 2023 reached 2,129.97 kcal, above the national consumption adequacy standard. Daily protein consumption was 62.43 g per person, also above the adequacy standard (Fig. 5.2).

When viewed by urban–rural classification, average daily consumption of calories per capita in Indonesia's urban areas is already above the national consumption adequacy standard that is 2,166.43 kcal, while in rural village areas the figure has also exceeded the requirement of 2,129.97 kcal. In both urban and rural areas, average daily protein consumption per capita is also above the minimum requirement. Daily consumption of protein per capita in urban areas is higher than in rural village areas, namely 64.42 g in urban areas compared to 59.23 g in rural areas.

The second agenda of the SDGs Desa is to end all forms of hunger in villages by 2030 and strive to create food security to ensure that everyone enjoys good food security, leading to a healthy life. Achieving this goal requires improving access to food and increasing agricultural production in a sustainable manner, which includes raising productivity and farmers' incomes, developing technology and market access, sustainable food production systems, and adding value to agricultural production.

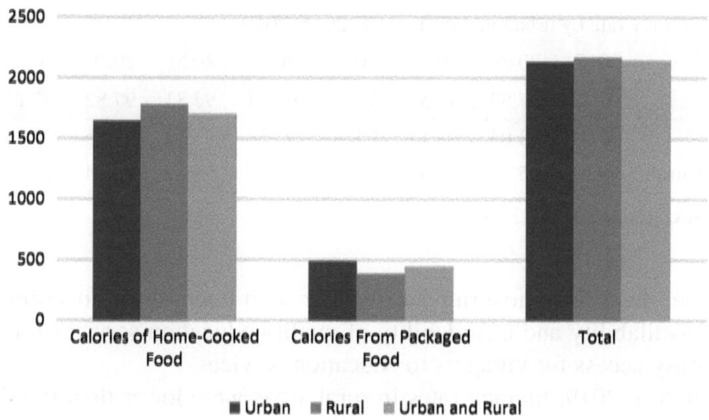

Fig. 5.2 Average daily calorie consumption per capita. *Source* Statistics Indonesia 2022a, b, c, d

5.6 SDGs Desa 3: Healthy and Prosperous Village

This goal seeks to ensure the good health of villagers, which is essential for the realization of prosperity. This goal requires that villagers have easy access to health services.

Indonesia's 2018 Basic Health Research (Riskesdas) examined ease of access to hospitals, considering mode of transportation, travel time, and cost. It found that 37.1% of households rated access to hospitals as easy, 36.9% as difficult, and 26% as very difficult.

For this reason, in order to achieve SDGs Desa 3, the village-level and upper-level governments must ensure access to health services for all villagers, affordable health insurance for villagers, lower maternal and infant mortality rates, greater provision of complete immunization to infants, higher use of contraceptives, as well as control of HIV/AIDS, tuberculosis, obesity, malaria, leprosy, filariasis (elephantiasis), control of drug abuse, and reduction of the birth rate among adolescents.

5.7 SDGs Desa 4: Quality Village Education

Development seeks to increase the income of the poor, ensure villagers' access to basic services, and protect all villagers from all forms of disaster. To achieve the goal of increasing income for the poor in villages, the main target of this goal is to improve the quality of village human resources. Education is a form of investment that determines the future of the nation. Education is essential for improving the quality and competitiveness of village human resources.

Table 5.3 Literacy rate by urban and rural areas, 2015–2019

Area Type	2015	2016	2017	2018	2019	2020	2021	2022	2023
Urban	97.43	97.53	97.57	97.56	97.71	97.81	97.82	97.91	97.94
Rural	92.91	93.03	93.11	93.30	93.56	93.64	93.65	94.21	94.54
Urban and Rural	95.22	95.38	95.50	95.66	95.90	96.00	96.04	96.35	96.53

Source Statistics Indonesia 2023a, b, c

Therefore, the village government together with upper-level governments must ensure the availability and affordability of quality education services for villagers, as well as easy access for villagers to education services.

From 2015 to 2019, literacy rates in rural areas were lower than in urban areas (Statistics Indonesia, 2019c). In 2019, the literacy rate in rural areas was 93.56%, while in urban areas it was 97.71%. Therefore, the achievement of SDGs Desa 4 is crucial toward achieving the SDGs (Table 5.3).

In addition, the average length of schooling in Indonesia remains very low, at 9.13 years in 2023, increasing slightly to 7.91 years in 2023. This means the average Indonesian has not completed senior high school or continued to senior high school.

Therefore, to achieve SDGs Desa 4, the village government, with support from upper government levels, must ensure villagers have access to accredited, quality, and affordable education services, as well as Islamic boarding schools.

This goal also focuses on the availability of skills-based education services for villagers, pre-school education services, non-formal education, as well as the availability of reading parks or village libraries.

5.8 SDGs Desa 5: Involvement of Village Women

With this goal, the village government, supported by various parties, is at the forefront of gender mainstreaming. By 2030, SDGs Desa 5 aims to create conditions that place all villagers in an equal position in all aspects of life, without discrimination against women.

In addition to being entitled to fair treatment, village women should also have equal opportunities in public affairs. Moreover, the achievement of SDGs Desa 5 requires the elimination of all forms of violence against women and children.

Among the indicators for achieving SDGs Desa 5 are: the availability of space and opportunities for the involvement of women in village governance, both as village officials and in the Village Council; median age at first marriage for women; health services for women; education services for women; and women's involvement in village development planning and implementation.

5.9 SDGs Desa 6: Village with Clean Water and Sanitation

Clean water and proper sanitation are basic human needs. Fulfillment of clean water and proper sanitation is still a problem worldwide. For this reason, the fulfillment of the need for clean water, drinking water, and sanitation is one of the SDGs.

From 2015 to 2019, there was an increase in the percentage of Indonesian households with access to safe drinking water, both in urban and rural village areas (Table 5.4).

Similarly, the percentage of households with access to proper sanitation has also increased in both urban and rural village areas from 2015 to 2023, as shown in the following Table 5.5.

The achievement of SDGs Desa 6 can be measured by several indicators, such as: household access to drinking water and proper sanitation reaching 100% by 2030; efficient use of drinking water; and actions to protect and restore ecosystems related to water resources, such as mountains, forests, wetlands, rivers, groundwater, and lakes.

5.9.1 SDGs Desa 7: Clean and Renewable Energy Village

Oil production in Indonesia is declining, from 346 million barrels (949,000 bpd) in 2009 to around 283 million barrels (778,000 bpd) in 2018 (National Energy Council, 2019). Among the reasons for this decline are that Indonesia's aging oil wells are becoming depleted, while investment in exploration for new wells is falling. Yet demand for energy is increasing amid ongoing population growth.

Table 5.4 Percentage of households with access to safe drinking water, 2015–2023

Area Type	2015	2016	2017	2018	2019	2020	2021	2022	2023
Urban	75.67	80.16	80.27	80.48	82.27	83.66	83.58	83.8	84.15
Rural	47.38	53.57	53.15	55.74	71.17	74.27	75.95	76.99	79.85
Urban and Rural	61.57	67.2	67.54	69.27	77.39	79.53	80.29	80.92	82.36

Source Statistics Indonesia 2023a, b, c

Table 5.5 Percentage of households with access to adequate sanitation, 2015–2023

Area Type	2015	2016	2017	2018	2019	2020	2021	2022	2023
Urban	81.30	81.05	80.82	81.55	95.63	96.08	96.00	95.51	96.02
Rural	60.58	60.72	62.10	64.18	81.15	82.74	83.91	84.93	85.70
Urban and Rural	70.97	71.14	72.04	73.68	89.27	90.21	90.78	91.05	91.72

Source National Socio-Economic Survey by Satistics Indonesia 2023a, b, c

Upgrading infrastructure and supporting technologies to provide cleaner and more efficient energy in all countries can fuel positive growth and help reduce environmental impacts (Institute for Essential Services Reform, 2019).

Energy is an economic driver and a prerequisite for development. Availability of sufficient and affordable energy can foster the fulfillment of basic needs such as food, health services, and quality education, thereby supporting quality human development. The provision of wider access to energy in remote areas and villages, where energy access is very limited, has contributed to a significant increase in community incomes (Mursanti and Tumiwa, 2019).

The electrification ratio is an indicator used by the government to measure the reach of energy supply in Indonesia. This ratio is determined by the number of houses connected to electricity, regardless of the quality of the electricity supply received. Not all villages in Indonesia have electricity. Data from the Ministry of Villages, Development of Disadvantaged Regions, and Transmigration shows that, as of 2023, 1,667 villages in Indonesia covering 258,252 families had no electricity at all (Fig. 5.3).

State electricity company PLN, in its Electricity Supply Business Plan, has continued providing village electricity in accordance with its *2017–2021 Village Electricity RoadMap*, as part of efforts to achieve its electrification target of 99.7% by 2025.

The village electricity strategy aims to expand the existing distribution network and build renewable energy power plants and hybrid power plants for very remote villages and use fossil-fueled power plants (PLN, 2017). As outlined in Table 5.6, electrification programs in 2019 targeted 3,141 villages, through both the Regular Village Electricity Program and the '2510 Program', the latter being a program that sought to provide electricity to 2,510 villages.

Therefore, to meet the population's need for energy, as well as to reduce greenhouse gas emissions, as per global commitments, the government continuously encourages development of new and renewable energy, as part of efforts to maintain energy security and independence. Under Presidential Regulation No. 79 of 2014 on the National Energy Policy, the target for new and renewable sources in the nation's energy mix is at least 23% in 2025 and 31% in 2050.

Indonesia has sufficient potential for new and renewable energy to achieve its primary energy mix target, as shown in Table 5.7.

The main objective of SDGs Desa 7 is to ensure that everyone has access to renewable energy. The achievement of this goal by 2030 can be measured by several indicators, including: household electricity consumption in the village reaches a minimum of 1,200 kWh; village households use gas or waste wood for cooking; and the use of renewable energy in the village.

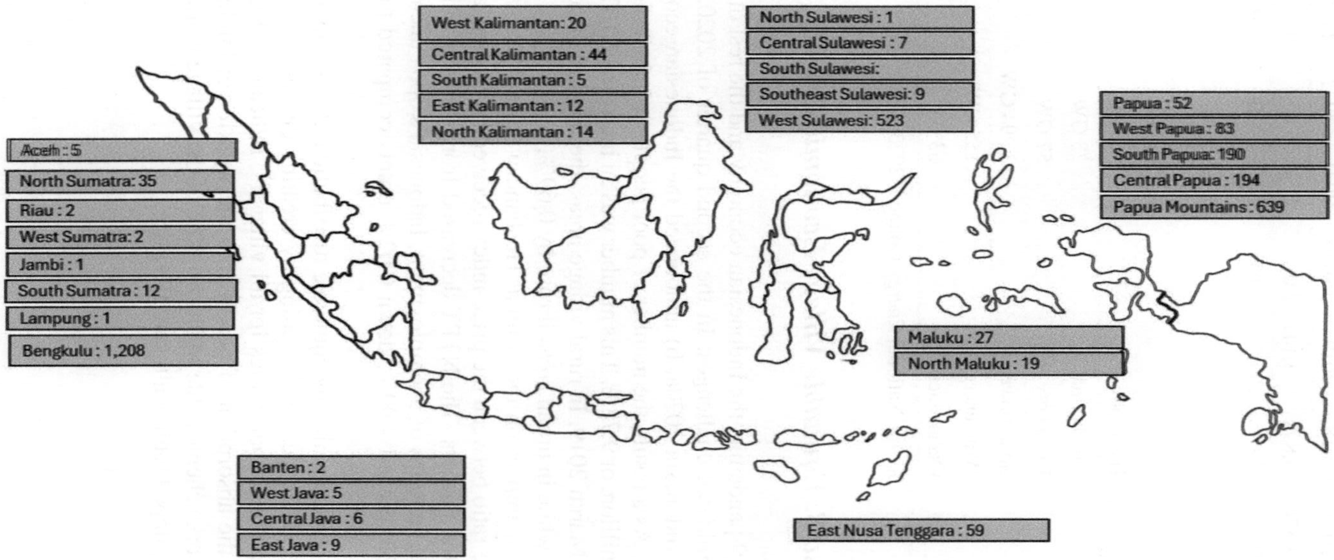

Fig. 5.3 Number of Indonesian Villages without electricity. *Source* Compiled from the Ministry of Villages, Development of Disadvantaged Regions, and Transmigration, 2024

Table 5.6 Village electricity program targets, 2016–2024

Program	2016	2017	2018	2019	2020	2021	2022	2023	2024
Villages	710	3,574	3,898	3,141	0	0	0	0	0
Percentage	98	98.5	99	99,5	100	100	100	100	100

Source PLN, RPJMN 2020–2024

Table 5.7 National renewable energy potential

Hydropower	95 GW
Geothermal	23 GW
Bioenergy	57 GW
Solar power	3,294 GW
Wind energy	155 GW
Ocean energy	63 GW

Source National Energy Council, 2023

5.9.2 SDGs Desa 8: Equitable Village Economic Growth

Facing the COVID-19 pandemic, the Indonesian economy and the rest of the world are dealing with formidable challenges. In the second quarter of 2020, Statistics Indonesia (Statistics Indonesia 2020a, b) announced the Indonesian economy had contracted by 5.32%. As a result, the number of poor people in Indonesia in March 2020 reached 26.42 million, or 9.78%. This number was an increase of 0.37% or 1.28 million people from March 2019. In rural village areas, the number of poor people decreased by 0.03%, while in urban areas it rose by 0.69%.

There was a similar impact on the Animal Husbandry Farmer Terms of Trade (NTPT), which is the ratio between the price index received by farmers against the price index paid by the farmers. The NTPT decreased from 99.94 in July 2020 to 98.64 in August 2020. In terms of unemployment, Indonesia's open unemployment rate in February 2020 reached 4.99%, down 0.02% points compared to February 2019.

Economic growth accompanied by equitable distribution of development results is the main target of SDGs Desa 8, including by creating decent jobs, as well as opening up new economic opportunities for all villagers. Indicators of the success of this goal include the absorption of people into the workforce; implementation of village cash-for-work schemes able to absorb 50% of the village workforce; and secure workplaces equipped with healthcare facilities.

5.9.3 SDGs Desa 9: Needs-Based Village Infrastructure and Innovation

Physical capital and human resources play important roles in economic growth. The availability of physical capital is closely linked to the availability of investment funds (Maryaningsih et al., 2014). A World Bank study (1994) states that the elasticity of gross domestic product (GDP) to infrastructure investment in a country is between 0.07 and 0.44. This means that a 1% increase in the availability of infrastructure will cause significant GDP growth of 7% to 44%.

Therefore, to achieve positive economic growth, the reliability of village infrastructure is crucial. This includes village roads, axis roads, as well as other infrastructure that supports the economic activities of villagers, such as infrastructure in agriculture, fisheries, and other sectors.

In addition to infrastructure, SDGs Desa 9 promotes the birth of village innovation in all fields, such as the economy and public services, and the development of superior village products. Therefore, SDGs Desa 9 is measured by several indicators of success that can be achieved by village-level and upper-level governments, including: reliable road conditions, piers and boat moorings, industrial growth in the village, and the contribution of industry to rural village economic growth.

5.9.4 SDGs Desa 10: Village Without Gaps

The issue of inequality always accompanies development activities. Income and wealth inequality have increased throughout much of the world in recent decades. In developing countries, inequality has increased by 11% in recent years. Currently, the richest 1% of the world's population controls up to 44.8% of global wealth, while the poorest 10% earn only 2% to 7% of global income (Oxfam, 2020).

Centralized development planning, which does not take into account spatial conditions, is more likely to exacerbate disparities, both between regions and between people. The gap is an indicator that the results of the economic development of a region or country are not enjoyed equally by various groups. Therefore, efforts to reduce inequality become an important component of overcoming poverty (Yusuf and Summer, 2015).

In March 2023, the level of inequality in the expenditure of the Indonesian population, as measured by the Gini Ratio, was 0.388 (Statistics Indonesia 2024a, b). This was a very slight increase from the November 2022 Gini Ratio of 0.381 and 0.381 in July 2022. The Gini Ratio in urban areas in March 2023 was at 0.409, up from 0.402 in November 2022 and 0,403 in July 2022. Meanwhile, the Gini Ratio in rural areas in March 2023 was 0.313, unchanged from November 2022, and slightly fall from 0.315 in July 2022 (Fig. 5.4).

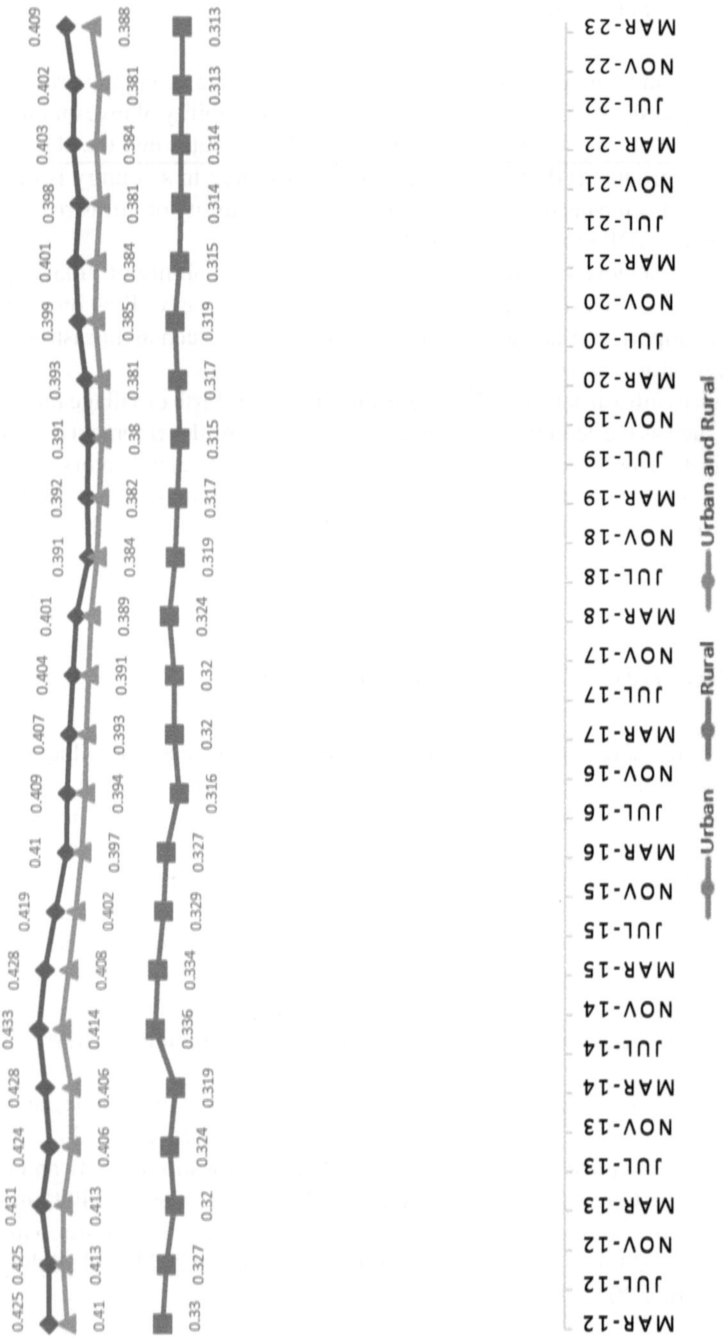

Fig. 5.4 Urban and rural Gini ratio, 2012–2023. *Source* Statistics Indonesia 2024a, b

Table 5.8 Percentage of adequate housing in urban and rural areas, 2015–2023

Tahun	Urban (%)	Rural (%)
2015	97,68	87,99
2016	98,19	89,59
2017	98,47	91,18
2018	98,55	92,26
2019	61,09	50,67
2020	63,24	54,82
2021	64,65	55,95
2022	63,45	565,84
2023	65,47	59,91

Source Statistics Indonesia 2024a, b

SDGs Desa 10 aims to reduce and eliminate income and wealth gaps by 2030. The success of achieving this goal is measured by the village Gini coefficient, village poverty rates, village development status, and the village civil liberties index.

5.9.5 *SDGs Desa 11: Safe and Comfortable Village Residential Areas*

Safe and affordable housing is a basic human need that must be met, so that people can conduct social and economic functions. Fulfillment of the right to housing requires adequate, clean, safe, and sustainable housing. The percentage of adequate housing in urban areas is higher than in rural areas (Table 5.8).

The need for adequate housing at affordable prices is often not matched by the availability of residential areas that meet the required standards, such as green open spaces, sports fields, places of business and trade, public facilities, sanitation, clean water, and waste management.

This SDGs Desa, to be achieved by 2030, targets the realization of an inclusive, safe, strong, and sustainable village, with several targets for achieving clean and healthy residential areas, creating environmental security through community self-help, as well as building the participation of all parties in village development.

5.9.6 *SDGs Desa 12: Environmentally Conscious Village Consumption and Production*

This goal involves efforts to reduce the environmental impact of human activity on the planet through appropriate production and consumption patterns. While economic growth is essential for improving people's welfare, it needs to be generated through

Table 5.9 Most common ways of handling garbage in rural areas, 2017

The handling way	Amount of villages	Percentage (%)
In the hole or burned	51,238	68.45
Other	4,694	6.27
River Irrigation Canal Sea Lake Got Sewer	4,588	6.13
The garbage can is then transported	14,332	19.15

Source https://idm.kemendesa.go.id/

sustainable development. It is therefore necessary to reduce our ecological footprint by changing the way we produce and consume goods and other resources. This focus is one of the SDGs Desa.

Efficiency in the management of shared natural resources, as well as efforts to reduce toxic waste and pollutants, are important targets for achieving this goal. Encouraging citizens, businesses, and consumers to recycle and reduce waste is also important. A shift in production and consumption activities is needed to increase sustainability.

Village policies need to be conducive to environmental conservation. One such policy is proper management of waste and rubbish. Handling waste properly supports the creation of sustainable production and consumption patterns.

Waste management is more common in urban areas than in rural village areas. In urban areas, waste disposal sites have been built; however, this is not the case in rural areas. Table 5.9 shows how urban households handle garbage (Indonesian Ministry of Villages, Development of Disadvantaged Regions and Transmigration 2015).

Indicators of the success of SDGs Desa 12 include village policies that regulate waste management in the business sector, efficient use of natural resources, and the presence of household and business waste management businesses.

5.9.7 SDGs Desa 13: Climate Change Responsive Village

The negative impacts of climate change have been felt by all countries in the world. According to the United Nations Development Programme (UNDP, 2019), greenhouse gas emissions are now more than 50% higher than in 1990. Average annual economic losses from climate-related disasters are in the hundreds of billions of dollars. Moreover, geophysical disasters, which are 91% climate-related, between 1998 and 2017 killed 1.3 million people and left 4.4 billion injured.

In Indonesia, greenhouse gas emissions nearly tripled between 1990 and 2015 (an increase of 196%), and the rate of increase is expected to grow further over the coming decade. The highest sources of emissions are deforestation and peat forest fires, followed by burning fossil fuels for energy. In addition, Indonesia recorded an 18% increase in carbon dioxide emissions during 2012–2017 due to increased emissions from power plants, industry, and the transport sector. Indonesia's total

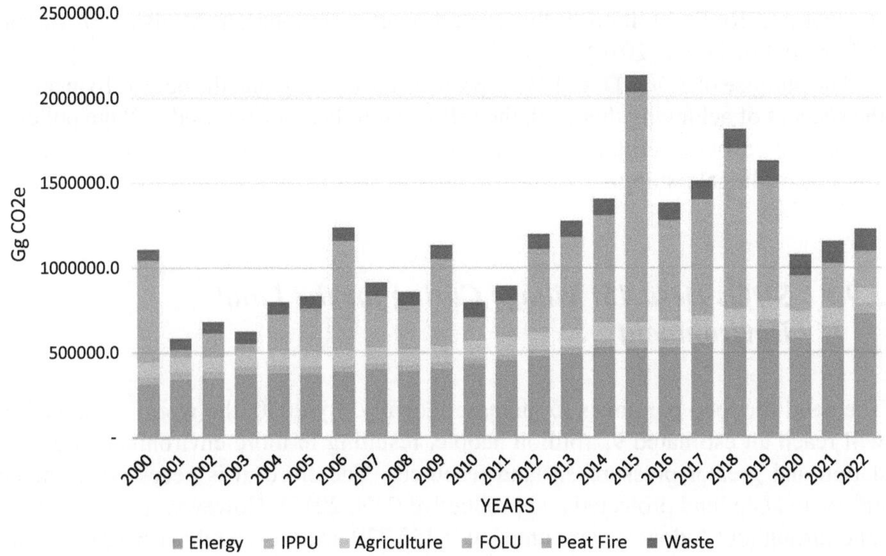

Fig. 5.5 Indonesia's total greenhouse gas emissions, 2000–2016. *Source* Ministry of Environment and Forestry, 2024

emissions vary widely from year to year, mostly as a result of peatland fires, as shown in Fig. 5.5.

SDGs Desa 13 aims to contribute to reducing the impact of global climate change, with several programs that can be carried out by village governments in accordance with their main tasks and functions. The success of achieving this goal can be seen from several indicators, one of which is the village disaster risk index.

5.9.8 SDGs Desa 14: Village Caring for the Marine Environment

More than 3 billion of the world's people depend on marine and coastal biodiversity for their livelihoods, contributing an estimated $28 trillion to the global economy each year (UN, 2019). In addition, the oceans absorb about 30% of the carbon dioxide produced by humans. Yet much of the ocean is heavily affected by pollution, with an average of 13,000 pieces of plastic debris for every square kilometer, while ocean acidification is rising, and 30% of the world's fish stocks are overexploited, falling below the level at which they can produce sustainable yields (Bappenas, 2019).

With its long coastline of 99,093 km and a sea area of 6.3 million km^2, Indonesia is a global center of marine biodiversity. The country's marine wealth provides economic and social benefits for the Indonesian people. In the first quarter of 2016, the Indonesian fisheries sector contributed Rp76.7 billion to the nation's gross domestic product and Rp77.7 billion in the second quarter (Indonesian Ministry of Marine Affairs and Fisheries, 2016).

The purpose of SDGs Desa 14 is to protect coastal areas and the ocean. To measure the success of achieving this goal, the following indicators are used: village policies on marine resource protection, a reasonable increase in fishing catches, and the absence of illegal fishing.

5.9.9 SDGs Desa 15: Village Caring for the Land Environment

The need for food continues to increase globally. By 2050, the world's population will reach an estimated 9.7 billion people, resulting in more environmental degradation and greater pressure on land resources. Therefore, it is necessary to make efforts to keep land protected and productive (UN, 2015). However, in 2018 critical land throughout Indonesia amounted to 9,453,729 hectares, while very critical land reached 4,552,721 hectares (Statistics Indonesia 2020a, b).

In order to fulfill current and future human needs, as well as to protect wildlife, SDGs Desa 15 aims to protect natural resources and wildlife. Indicators used to measure the success of this goal include village government policies related to efforts to conserve biodiversity, the amount of green open spaces, and endangered species numbers (Table 5.10).

5.9.10 SDGs Desa 16: Village of Peace and Justice

This goal aims to create peaceful and secure village conditions to ensure that village governments can work fairly and effectively. Therefore, several efforts must be made by the village government and upper levels of government, namely to significantly reduce all forms of violence and to find long-term solutions to village conflicts. So far, the incidence of village crime is still quite high (Fig. 5.6).

During the 2020–2023 period, theft was the most common type of crime in villages and urban communities in Indonesia, occurring in 36–45% of all villages.

Meanwhile, the incidence of mass fights has increased, as shown in Fig. 5.7.

Therefore, SDGs Desa 16 sets several targets to be achieved by 2030: eradication of crime, fighting, domestic violence, and violence against children; preservation of the culture of *gotong royong* (mutual assistance) in the village; increasing the village democracy index; and eradication of human trafficking and child labor.

Table 5.10 Critical land area levels (ha) by province, 2018

	Area and distribution of critical land by province (Hare)		
	Critical	Very critical	Total
ACEH	126,238	190,399	316,637
NORTH SUMATRA	836,530	502,280	1,338,810
WEST SUMATRA	174,748	477,221	651,970
RIAU	435,709	275,164	710,873
JAMBI	111,855	102,130	213,985
SOUTH SUMATRA	602,938	130,818	733,756
BENGKULU	120,598	28,289	148,887
LAMPUNG	381,933	21,977	403,910
BANGKA BELITUNG DISTRICT	20,426	261	20,687
RIAU DISTRICT	4,269	3,961	8,230
WEST JAVA	231,445	679,748	911,192
CENTRAL JAVA	158,374	217,359	375,733
IN YOGYAKARTA	78,333	790	79,123
EAST JAVA	86,937	345,289	432,225
BANTEN	236,325	94,083	330,407
BALI	44,611	2,284	46,895
WEST SOUTHEAST NUSA	55,158	10,641	65,799
EAST EAST NUSA	831,330	9,585	840,914
WEST KALIMANTAN	996,541	19,090	1,015,631
CENTRAL KALIMANTAN	768,756	92,484	861,240
SOUTH KALIMANTAN	286,459	225,135	511,594
EAST KALIMANTAN	156,839	118,433	275,272
NORTH KALIMANTAN	163,520	36,214	199,734
NORTH SULAWESI	197,748	92,034	289,782
CENTRAL SULAWESI	164,305	100,569	264,874
SOUTH SULAWESI	155,587	294,019	449,606
SOUTHEAST SULAWESI	409,624	15,031	424,655
GORONTALO	73,443	258,855	332,298
WEST SULAWESI	74,085	14,336	88,421
MALUKU	299,577	30	299,607
NORTH MALUKU	372,940	14,949	387,889
WEST PAPUA	427,504	9,784	437,288
PAPUA	369,042	169,481	538,523
INDONESIA	9,453,729	4,552,721	14,006,450

Source Statistics Indonesia 2022b

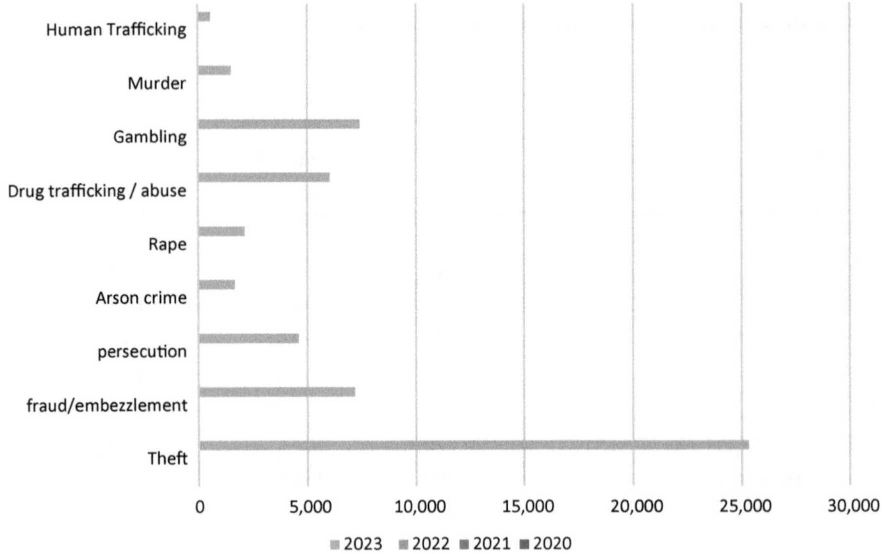

Fig. 5.6 Types of crime by incidence in villages in 2020, 2021, 2022, and 2023. *Source* https://www.bps.go.id/

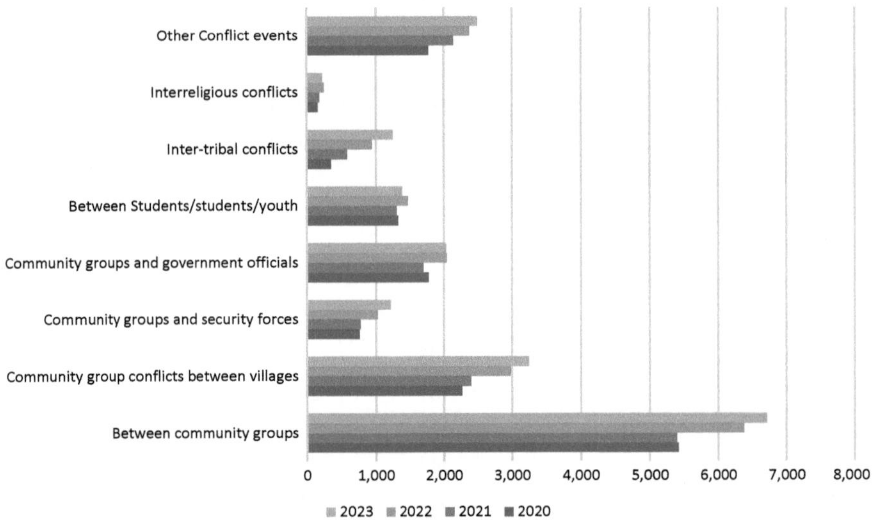

Fig. 5.7 Types of mass conflict by incidence in villages in 2020, 2021, 2022, and 2023. *Source* https://idm.kemendesa.go.id/

5.9.11 SDGs Desa 17: Partnerships for Village Development

This goal is a means of implementing and revitalizing village partnerships to realize all development goals. Village development will only succeed optimally with the involvement of all related parties. Starting from community leaders, youth village change-makers, women driving the village economy, universities, the business sector, upper levels of government, village officials, and the Village Council.

The achievement of sustainable development requires good village partnerships and cooperation, including with other villages and subdistricts. Therefore, there should be no feelings of superiority between villages if they are to achieve the SDGs Desa.

All sectors need to be improved by leveraging technology and knowledge to create innovation. Every village needs to implement an organized and coordinated policy, especially with the upper-level governments, universities, and the business sector.

The achievement of this goal is measured by several indicators, including the existence and form of village cooperation with third parties, availability of internet access in the village, village statistics, village trade data, and other village activities.

5.9.12 SDGs Desa 18: Dynamic Village Institutions and Adaptive Village Culture

The institutions of a village are a vital component of sustainable development. For this reason, village institutions must be strengthened so they can support the achievement of the SDGs.

SDGs Desa 18 strives to maintain local wisdom, as well as revitalize and mobilize all elements of village-level institutions. This is because the involvement of strong and functioning village institutions in people's lives will support diversity in a dynamic village, as well as drive the achievement of the SDGs Desa.

In 2018, the rate of public participation in local social activities amounted to 85.43% (Statistics Indonesia 2018a, b). If categorized by area of residence, more rural residents participate in social activities than urban residents (88.56% compared to 82.94%). The three social activities most participated in by people aged 10 years and over in a three-month period were funerals and memorial ceremonies (63.87%), religious activities (61.87%), and other social activities (46.82%) (Table 5.11).

To measure the achievement of SDGs Desa 18, several indicators are used, including: sustainability of mutual assistance and mutual cooperation activities, participation of religious leaders in village development activities, protection of the weak and orphans, preservation of village culture, and resolving residents' problems through cultural approaches.

Table 5.11 Participation rates (%) in social activities in urban and rural areas, 2021

Area type	Social activities	Religious activities	Practical activities	Sport/ games	Mutual assistance	Area Type	Social activities
Urban	43.6	0.58	6.89	12.69	52.93	30.85	20.51
Rural	54.67	1.07	6.62	14.77	65.08	40.67	26.5
Total	48.32	0.79	6.78	13.58	58.11	35.03	23.06

Source Sociocultural and Education Module of National Socio-Economic Survey, Statistics Indonesia 2021

References

Concern Worldwide and Welthungerhilfe. 2019. *Global Hunger Index: The Challenge of Hunger And Climate Change*. Dublin: Helvetas.

Fan, S., and P. Polman. 2014. An Ambitious Development Goal: Ending Hunger and Undernutrition by 2025. In *Global Food Policy Report*. Washington, D.C.: IFPRI.

Food and Agriculture Organization (FAO). 2008. *An Introduction to the Basic Concepts of Food Security*. Rome: FAO.

Indonesian Ministry of Villages, Development of Disadvantaged Regions, and Transmigration. 2015. *Developing Villages Index*. Jakarta: Ministry of Villages, Development of Disadvantaged Regions, and Transmigration.

Indonesian Ministry of Marine Affairs and Fisheries. 2016. Kelautan dan Perikanan dalam Angka Tahun 2016 (Marine and Fisheries in Figures 2016. Jakarta: Indonesian Ministry of Marine Affairs and Fisheries.

Mursanti, E., and F. Tumiwa. 2019. *Strategi Penyediaan Akses Listrik di Perdesaan dan Daerah Terpencil di Indonesia*. Jakarta: Institute for Essential Services Reform.

Maryaningsih, N., O. Hermansyah, and M. Savitri. 2014. Pengaruh Infrastruktur terhadap Pertumbuhan Ekonomi Indonesia. In *Buletin Ekonomi Moneter dan Perbankan*, Vol. 17, No. 1.

National Development Planning Agency (Bappenas). 2019. Tujuan 14: *Ekosistem Lautan*. Jakarta: Bappenas.

Oxfam. 2020. World's Billionaires Have More Wealth than 4.6 Billion People. Accessed from: www.oxfamamerica.org/press/worlds-billionaires-have-more-wealth-than-46-billion-people.

State Electricity Company (PLN). 2017. Jakarta: PLN

Statistics Indonesia (BPS). 2018a. *Survei Sosial Ekonomi Nasional (Susenas) Tahun 2018*. Jakarta: BPS

Statistics Indonesia (BPS). 2018b. *Statistik Kesejahteraan Rakyat Tahun 2018*. BPS. Jakarta: BPS.

Statistics Indonesia (BPS). 2020a. *Statistik Kriminal Tahun 2019*. BPS. Jakarta: BPS.

Statistics Indonesia (BPS). 2020b. *Tingkat Ketimpangan Pengeluaran Penduduk Indonesia Maret 2024*. Jakarta: BPS

Statistics Indonesia (BPS). 2021. *Susenas MSBP Tahun 2021*. BPS. Jakarta: BPS.

Statistics Indonesia (BPS). 2022a. *Laporan Bulanan Data Sosial Ekonomi*, Edisi 124, September 2020. Jakarta: BPS.

Statistics Indonesia (BPS). 2022b. *Luas Lahan Kritis menurut Provinsi dan tingkat Kekritisan*. Accessed from: www.bps.go.id/indicator/60/588/1/luas-lahan-kritis-menurut-provinsi-dan-tingkat-kekritisan-lahan.html.

Statitics Indonesia (BPS). 2022c. *Profil Kemiskinan di Indonesia September 2021*. Jakarta: BPS

Statistics Indonesia (BPS). 2022d. *Statistik Indonesia Tahun 2020*. Jakarta: BPS.

Statistics Indonesia (BPS). 2022e. *Tingkat Ketimpangan Pengeluaran Penduduk Indonesia 2020*. Jakarta: BPS

Statitics Indonesia (BPS). 2023a. *Profil Kemiskinan di Indonesia September 2022*. Jakarta: BPS

Statitics Indonesia (BPS). 2023b. *Profil Kemiskinan di Indonesia Maret 2023*. Jakarta: BPS

Statistics Indonesia (BPS). 2023c. *Survei Sosial Ekonomi Nasional (Susenas) Tahun 2023*. Jakarta: BPS.

Statitics Indonesia (BPS). 2024a. *Profil Kemiskinan di Indonesia Maret 2024*. Jakarta: BPS

Statitics Indonesia (BPS). 2024b. *Tingkat Ketimpangan Pengeluaran Penduduk Indonesia Maret 2024*. Jakarta: BPS

United Nations. 2015. *Transforming Our World: The 2030 Agenda For Sustainable Development*. New York: UN.

UNDP. 2019. *Goal 13: Climate Actions*. New York: UNDP.

World Bank. 1994. *World Development Report: Infrastructure for Development*. New York: Oxford University Press.

Yusuf, A.A., and A. Sumner. 2015. Growth, Poverty, and Inequality under Jokowi. In *Bulletin of Indonesian Economic Studies*, No. 51.

Chapter 6
SDGs Desa 18: Dynamic Village Institutions and Adaptive Village Culture

Abstract At the conceptual level, it is easy to define development as a process of change for society; however, experiencing life amid the vortex of such change can be a formidable challenge. Development does not always grow as the fruit of internal community dynamics, but may also be transplanted from outside the village.

At the conceptual level, it is easy to define development as a process of change for society; however, experiencing life amid the vortex of such change can be a formidable challenge. Development does not always grow as the fruit of internal community dynamics, but may also be transplanted from outside the village. Development therefore requires the community's adaptive capacity to accept change (Suwarsono and So, 2014).

Not surprisingly, some village residents have criticized problems that arise from living amid ongoing development (Berger, 2005; Husaini et al., 2011; Israel, 1992). Much of this criticism tends to come from the most vulnerable or disadvantaged residents, who find it difficult to follow the steps of development. Criticism includes development processes that leave residents behind, silence residents' voices, cause physical or social exclusion, and even result in deaths among villagers.

One strategy to eliminate feelings of alienation from village development is to bring the whole process closer to the unique characteristics of each village community. It is fortunate that the village is seen as a distinct social entity, because interaction between its residents is intensive: they meet face to face almost every day, helping each other when they are faced with mutual difficulties.

So, it remains only to bring the development concepts, objectives, and implementation strategies more into line with the village's structure and typical daily behavior. This point was the reason for the formulation of SDGs Desa 18: Dynamic Village Institutions and Adaptive Village Culture.

Just don't develop

Make the worms happy

Iwan Fals, lyrics of Lancar

A. H. Iskandar, *Accelerating the Achievement of the National Sustainable Development Goals in Indonesia*, Engaging Indonesia, https://doi.org/10.1007/978-981-97-8919-1_6

6.1 Various Emergences of SDG 18

By synthesizing various global concepts of development, the global SDGs were compiled in 2015, consisting of 17 goals with 169 indicators (UN, 2015). However, once these had been declared, there were calls for additional indicators. Moreover, the 2030 deadline for achieving the SDGs means there is a long-term development span of 15 years, which must be able to adapt to socio-economic developments.

Recent changes include several low-income countries having grown to become middle-income countries, including Indonesia (Alisjahbana and Murniningtyas, 2018; Basile and Baud, 2019). Also, the economies of countries such as China, India, and Brazil are growing rapidly with great strength. While some countries enjoy economic development, others are suffering from conflict and war, prompting an increase in international migration. Unfortunately, the issue of international migration is also intertwined with the issue of global terrorism. Another developing issue is the increasing use of technology, from the internet to artificial intelligence. Telematics technology has also entered agricultural businesses in villages (Ragkos et al., 2019).

These new issues were not included in the SDGs declared in 2015. Rather, some of the new issues have been adopted as additional indicators for the SDGs. In 2020 alone, the number of indicators of the SDGs grew from 169 to 247 indicators, of which 231 are unique, while 12 exist across several of the SDGs.[1] It is planned that this kind of indicator adjustment will take place every year until 2025.

However, that alone is not enough. There are still those who need new development goals for indicators that are difficult to incorporate into the existing 17 SDGs. At first glance, SDG 17 incorporates some aspects of data and information, but it clearly lacks references to big data and the internet of things, which underlie the Fourth Industrial Revolution, also known as Industry 4.0 (Schwab, 2019). Therefore, some have called for the creation of SDG 18 to cover living with artificial intelligence, proposed as SDG 18: Life with Artificials.

There is also an academic proposal for SDG 18 to cover animal rights, formulated as SDG 18: Animal Health, Welfare and Rights (Visseren-Hamakers, 2020). This proposal is based on the consideration of pre-existing rights for non-human natural entities, such as animal rights in Germany; rights of nature in Ecuador, Colombia, India, and New Zealand; and river rights in New Zealand and India. The animal rights-oriented SDG 18 would aim to protect ecosystems and species, prevent hunting and the illegal extraction of marine resources, and fight invasions of alien species into natural ecosystems.

Although a lack of scope in the SDGs has been acknowledged, none of the proposals for SDG 18 to complement the existing goals have been formally accepted.

[1] Global indicator framework for the Sustainable Development Goals and targets of the 2030 Agenda for Sustainable Development. Accessed from https://unstats.un.org/sdgs/indicators/Global%20Indicator%20Framework%20after%202,020%20review_Eng.pdf on 31 January 2022.

Perhaps this is because the proponents are outside the government, namely community organizations and scientists. In addition, the proposals come from parties focused only on one or several related aspects, rather than all aspects of the SDGs.

This is what differentiates Indonesia's SDGs Desa 18: Dynamic Village Institutions and Adaptive Village Culture. Villages will implement all of the SDGs, so this huge social change requires formal policy support. The policy is directed at placing the SDGs into the village context. Once they fit the village context, the SDGs can be achieved more effectively.

In addition, SDGs Desa 18 was devised by the state, represented by the Ministry of Villages, Development of Disadvantaged Regions, and Transmigration. SDGs Desa 18 is part of the Regulation of the Minister of Villages, Disadvantaged Regions, and Transmigration No. 13 of 2020 on Priorities for Use of Village Funds in 2021. This means the decision is legal and binding for 75,265villages in Indonesia to implement the SDGs Desa in 2021. This obligation was extended under the Minister's Regulation No. 7 of 2021 on Priorities for Use of Village Funds in 2022.

6.2 The Urgency of SDGs Desa 18

When attempts are made to implement the global SDGs at the village level, their weaknesses immediately become apparent. The global narrative, supported by statistical uniformity across countries, has moved away from the sociologically diverse local context. This neglect of the local context means the implementation of the SDGs at the village level can be perceived as a hierarchical task, whose governance reaches all the way to the top of the UN leadership (Kanie et al., 2017). The absence of context reduces any sense of belonging among local residents, and in the long term can foster apathy toward village development (Cernea, ed. 1985; Korten, 2001).

The SDGs Desa are not merely wishful thinking, whimsical notions, or lip service. They are an official policy that must be implemented directly in the field, so villagers can soon feel the results and benefits. Regulation No. 13 of 2020 indicates the legality of the policy, which includes SDGs Desa 18 and began to be implemented with the preparation of plans for all villages in the archipelago in January 2020.

The SDGs Desa are not solely directed at achieving the global SDGs or international interests, but are chosen as a strategy to accelerate the wheels of village development itself. This is because various indicators of development goals have been accumulated in the SDGs, so it is sufficient to add indicators that are appropriate to the context of villages in Indonesia.

In Indonesia, the SDGs were not initially adapted to be localized to the village level. The localization obligation originally extended only to the provincial and regency levels, by preparing regional action plans, accompanied by a process of monitoring and evaluating the results and benefits.[2] Therefore, when localization entered the village realm, a new strategy needed to be adopted.

It can be said the localization of the SDGs into SDGs Desa did not take place from the top down, such as integrating local planning documents into the SDGs' indicators. The opposite approach was taken by adjusting the SDGs to suit the village context, which has so far provided specificity in development. As shown in the previous chapter, the Village Law provides an explanation of village development, which is then embodied in the annual and medium-term planning documents for villages. These plans have then been implemented, bringing the results and benefits of development. Village development programs have been designed and implemented in accordance with local wisdom, so that local customs and habits shape development processes, impacts, and benefits. For example, the tradition of *gotong royong* (mutual assistance) is adapted to be labor-intensive, thereby creating job opportunities in the village. Even if there is no labor-intensive work scheme, but road repairs must be done, *gotong royong* is again carried out by the village. This kind of adaptation can occur because of the daily interactions between villagers.

Wisdom like this, which helps villages adapt positively to change, underlies why the village-based context was needed to form SDGs Desa 18. The choice of this additional goal is interesting because it has many indicators and covers almost all aspects of life. It therefore can ensure village development is more holistic and easier to observe through measurable statistics, so that through careful direction, residents can receive the benefits more quickly.

The formulation and wording of SDGs Desa 18: Dynamic Village Institutions and Adaptive Village Culture, as well as a village development icon, makes it easier for villagers to understand development holistically. This village-specific goal is also more in line with the social and cultural conditions in the village.

6.3 Dynamic Village Institutions

One of the keys to implementing the SDGs Desa is dynamic institutions. Such institutions must be willing to follow the latest data, so they can be agile in formulating policies that match current needs. The importance of institutions that are able to support the implementation of the SDGs has been demonstrated in various locations (Alisjahbana and Murniningtyas, 2018; Biermann and Kanie, 2017; Crowther et al., 2018).

In the village, local government institutions play a vital role. The village head and his or her staff carry out various functions to maintain peace among residents amid change (Breman and Wiradi, 2004; Husken, 1998). At the same time, they become leaders, whose legitimacy is maintained through their service to citizens (Geertz,

[2] Regulation of the Minister of National Development Planning/Head of Bappenas No. 7 of 2018 on Coordination, Planning, Monitoring, Evaluation and Reporting on the Implementation of the Sustainable Development Goals.

1986). When citizens are satisfied with their services, their legitimacy increases. Thus, the village government can function to empower its citizens (Supriatna, 1997).

The highest institution for village-level decision-making is the village meeting. Implemented in a variety of ways, the meeting process always involves discussion through deliberation. This institution is a critical point for mobilizing change, including in the form of village development. Dialogue becomes an essential process to understand new ideas, concepts, and development indicators from outside the village. *Rencana Umum Penyediaan Tenaga Listrik 2017–2026.*

The various social institutions that emerged in the village reflect the needs of the villagers, because they were usually formed to meet the specific needs of citizens (Soekanto, 2001). Therefore, they must actively participate in discussing development designs, through involvement in village deliberations.

Once a proposal for progress is agreed upon at a meeting, annual and medium-term development plan are drawn up. Under regulations, the planning must refer to the latest village data. The village compiles its own annual data in the first semester, to be used in the second semester as the basis for designing village development plans for the following year. The village development planning process usually begins with the identification of deficiencies in services, facilities, and components of village progress. The village deliberations then generally decide to fulfill the village's needs that have not yet been met. This process is similar to the steps to fulfill the objectives of the SDGs Desa, so this provides an opportunity for the simultaneous implementation of the SDGs Desa.

Opportunities to maintain village economic resilience include the development of village-owned enterprises. That's because village deliberations enable village-owned enterprises to be directed to support residents' economic activities already in development, so they can develop even more rapidly. For example, in a village where a lot of people work in businesses making clothes, the village-owned enterprise develops a business to supply their products, especially to a wider market. Part of the profits from the village-owned enterprise are required to be a component of village-generated income. Once this village-generated income is recorded in the village budget, it can be used for village development.

In addition, village-generated income can also be used to increase village assets. Productive assets, such as land, then become a larger source of village-generated income. Such asset enlargement is expected to further expand services to villagers.

6.4 Adaptive Village Culture

Culture provides a sense of security, a sense of home, and a sense of comfort, especially when thinking and taking action (Stange, 2009). The SDGs Desa highlight the capacity of culture to be adaptive to new perspectives. The capacity to be adaptive encourages citizens to deal with outside entities, and opens the door for dialogue to different points of view. An adaptive culture also raises awareness of the latest data and information, thus opening opportunities to use them to build a more advanced village.

Following various studies on the influence of religion on sustainable development in Asia (Duc, 2017; Rubianto, 1996), it was only recently that SDGs Desa 18 was included as an indicator of achieving village-level development goals. Religion ignites high motivation for generosity, both in the construction of places of worship, as well as other mutual cooperation activities in the village.

Religiosity also makes it easier for residents to help the poor and orphans. That's because religions have a tradition of charity and alms for the needy. In Islam, for example, the Prophet, whose guidance is followed by adherents, was an orphan who preached about importance of helping the poor and orphans.

Religious values in the village are fostered by local religious leaders. They are in daily contact with residents, so they often participate in development planning, implementation, and evaluation. As community figures, they play an important role in conveying residents' voices to participants in village deliberations, or to village authorities. It is from this role of conveying aspirations that they gain respect.

Enduring traditions from within the village, which still influence opinions, behavior, and decisions, need to be preserved. This gives residents peace of mind as they go about their daily lives, because such traditions contain the values of goodness and truth (Dahlan, 1978; Santosa, 2011), including when the village changes as a consequence of development. In addition to being adapted to the local environment, development concepts from outside the village should also be adapted to the local language and customs.

As with the adaptation of religious values into daily life, customary wisdom is also adapted by traditional elders in the village deliberation process. This is where a traditional point of view is communicated in line with a development paradigm, so the new point of view makes the implementation of development activities feel in accordance with the pattern of village life.

A key indicator of the village's cultural strength is its ability to independently handle internal conflicts. This ability can be demonstrated by conflict resolution conducted by the village itself, not by powers outside the village, let alone by law enforcement officers.

This ability shows the unique strength of the archipelagic village context, which underpins the concept of SDGs Desa 18: Dynamic Village Institutions and Adaptive Village Culture.

References

Alisjahbana, A. S., and E. Murniningtyas. 2018. *Tujuan Pembangunan Berkelanjutan di Indonesia: Konsep Target dan Strategi Implementasi*. Bandung: Unpad Press.

Basile, E., and I. Baud. 2019. Introduction: Reflecting on Development Studies in the New Millennium. In *Building Development Studies for the New Millennium*, edited by I. Baud, E. Basile, T. Kontinen and S.V. Itter. Cham, Switzerland: Palgrave Macmillan.

Berger, P.L. 2005. *Piramida Kurban Manusia: Etika Politik dan Perubahan Sosial*. Indonesian translation. Jakarta: LP3ES.

Biermann, F., and N. Kanie. 2017. Conclusion: Key Challenges for Global Governance through Goals. In *Governing through Goals: Sustainable Development Goals as Governance Innovation*, edited by N. Kanie, and F. Biermann. Massachusetts: MIT Press.

Breman, J., and G. Wiradi. 2004. *Masa Cerah dan Masa Suram di Pedesaan Jawa, Studi Kasus Dinamika Sosio-Ekonomi di Dua Desa Menjelang Akhir Abad Ke-20*. Indonesian translation. Jakarta: LP3ES.

Cernea, M. (ed.). 1985. *Putting People First: Sociological Variables in Rural Development*. New York: Oxford University Press.

Crowther, D., S. Seifi, and A. Moyeen. 2018. Responsibility and Governance in Achieving Sustainability. In *The Goals of Sustainable Development: Responsibility and Governance*, edited by D. Crowther, S. Seifi, and A. Moyeen. Switzerland: Springer.

Dahlan, M.A. 1978. *Sosialisasi Pola Hidup Sederhana. Prisma*, Vol. 7 No. 10.

Duc, A.L. 2017. Religious Environmentalism and Environmental Sustainability in Asia. In *Sustainable Development Goals in the Asian Context*, edited by J. Servaes. Singapore: Springer.

Geertz, C. 1986. *Mojokuto: Dinamika Sosial Sebuah Kota di Jawa*. Indonesian translation. Jakarta: Grafitpers.

Husaini, E.W. Widianto, and E.D. Kristanto. 2011. *Aksi Melawan Imperium Modal: Gerakan Rakyat Melawan Pabrik Semen*. Yogyakarta: Resist Book.

Husken, F. 1998. *Masyarakat Desa dalam Perubahan Zaman, Sejarah Diferensiasi Sosial di Jawa 1830–1980*. Indonesian translation from Village Society in Changing Times: A History of Social Differentiation in Java 1830–1980. Jakarta: Gramedia Widiasarana Indonesia.

Israel, A. 1992. *Pengembangan Kelembagaan Pengalaman Proyek-Proyek Bank Dunia*. Jakarta: LP3ES.

Kanie, N., S. Bernstein, F. Biermann, and P.M. Haas. 2017. Introduction: Global Governance through Goal Setting. In *Governing through Goals: Sustainable Development Goals as Governance Innovation*, edited by N. Kanie, and F. Biermann. Massachusetts: MIT Press.

Korten, D.C. 2001. *Menuju Abad Ke-21: Tindakan Sukarela dan Agenda Global*. Indonesian translation. Jakarta: Yayasan Obor Indonesia.

Ragkos, A., G. Koutouzidou, S. Koutsou, and D. Roustemis. 2019. A New Development Paradigm for Local Animal Breeds and the Role of Information and Communication Technologies. In *Innovative Approaches and Applications for Sustainable Rural Development: 8th International Conference, HAICTA 2017, Chania, Crete, Greece, September 21–24, 2017, Selected Papers*, edited by A. Theodoridis, A. Ragkos and M. Salampasis. Switzerland: Springer.

Rubianto, V. 1996. *Paradigma Asia: Pertautan Kemiskinan dan Kereligiusan dalam Teologi Aloysius Pieris*. Yogyakarta: Kanisius.

Santosa, I.B. 2011. *Laku Prihatin: Investasi menuju Sukses ala Manusia Jawa*. Yogyakarta: Memayu.

Schwab, K. 2019. *Revolusi Industri Keempat*. Indonesian translation. Jakarta: Gramedia.

Soekanto, S. 2001. *Sosiologi Suatu Pengantar*, 4th ed. Jakarta: Raja Grafindo Persada.

Stange, P. 2009. *Politik Perhatian, Rasa dalam Kebudayaan Jawa*. Yogyakarta: LKiS.

Supriatna, T. 1997. *Birokrasi, Pemberdayaan dan Pengentasan Kemiskinan*. Bandung: Humaniora Utama Press.

Suwarsono, A, and Y. So. 2014. *Perubahan Sosial dan Pembangunan*. Jakarta: LP3ES.

United Nations. 2015. *Transforming Our World: The 2030 Agenda For Sustainable Development*. New York: UN.

Visseren-Hamakers, I.J. 2020. *The 18th Sustainable Development Goal*. In *Earth System Governance*, No. 3.

Chapter 7
Epilogue: Villages of Hope

Abstract The SDGs Desa are an integrated effort presented as an alternative action to accelerate Indonesia's achievement of the Sustainable Development Goals at the village level.

The SDGs Desa are an integrated effort presented as an alternative action to accelerate Indonesia's achievement of the Sustainable Development Goals at the village level. At the national level, the establishment of the SDGs Desa will make a major contribution toward Indonesia's achievement of the global SDGs. The SDGs Desa support and accelerate the achievement of Presidential Regulation No. 59 of 2017 on the Implementation of the SDGs.

Presidential Regulation No. 59 of 2017 Presidential Regulation No. 59 of 2017 itself is a follow-up to the Indonesian government's approval of the United Nations' global development resolution, *Transforming Our World: The 2030 Agenda for Sustainable Development,* which included the global development agenda of 17 goals and 169 targets, known as the SDGs (UN, 2015).

The deadline for achieving the SDGs is 2030, and the goals are universal and inclusive in nature, with the promise of "leave no one behind". The SDGs are a global development agenda, a continuation of the Millennium Development Goals (MDGs) agreed upon by 189 member states of the UN at the Millennium Summit in September 2000.

All of these global development agendas are part of the international community's effort to correct development, which was previously viewed only as an economic phenomenon. The past focus on economic development had failed to sufficiently improve public welfare and instead created income disparities between communities and regions, and wealth gaps between nations. The MDGs and SDGs are efforts to correct the development concept which held that growth of gross national product

Indonesia won't be big because of the torch in Jakarta,

but Indonesia will shine because of the candles in the village.

Bung Hatta

A. H. Iskandar, *Accelerating the Achievement of the National Sustainable Development Goals in Indonesia*, Engaging Indonesia, https://doi.org/10.1007/978-981-97-8919-1_7

per capita would create a trickle-down effect to the wider community in the form of jobs and other economic opportunities.

This book has presented many arguments to show that by mainstreaming the SDGs Desa in development policies, villages will play a role and have enormous potential in supporting and accelerating the achievement of the global SDGs. Among these arguments is the existence of high social capital in the village, such as togetherness, mutual cooperation, solidarity, and also trust, all of which are important capital for development and achieving the SDGs Desa. Several countries have even managed to get through economic crises by optimizing the potential of the village. In Nigeria, for example, an economic crisis resulted in the dismissal of urban industrial workers. At that time, the Nigerian government was trying to encourage the acceleration of village development through the agricultural sector and succeeded in opening up employment opportunities, as well as increasing the country's economic growth (Yustika, 2016).

Another argument is the policy of making villages the spearhead for the development of a national food barn. This is expected to help achieve national food security, so that nutrition will be improved. Also, one of the main challenges faced by the village is the improvement of the nutrition of rural communities. For this reason, future village development must refer to the objectives of the SDGs Desa. Village development must have an impact on the villages, which can then contribute to development.

The principle of "leave no one behind" in the SDGs refers to the state's obligation to ensure that none of its citizens are left behind, forgotten, or marginalized from their rights in development. In this context, citizens who live in remote and outer areas, such as villages, and are vulnerable groups – such as the poor, indigenous people, women, children, farm laborers, people with disabilities, the elderly – must be prioritized and reached first.

In addition, this book has noted the SDGs Desa can contribute to 74 percent of the achievement of national sustainable development. In other words, if village development is focused on efforts to support the achievement of the SDGs, it can be concluded that 74 percent of the SDG targets will be reached because of contributions from the nation's villages.

This calculation is based on territorial and civic aspects. From the territorial aspect, Ministry of Home Affair has classified 91 percent of Indonesia's territory as village areas and as many as 11 of the SDGs (Goals 7–17) are closely related to village areas. In terms of the civic aspect, 43 percent of Indonesians live in rural village areas (BPS, 2020a, 2020b), and at least six of the SDGs (Goals 1–6) are closely related to villagers.

These arguments strengthen the position of the village as the main axis in achieving the SDGs. Although there is much pessimism about the readiness and ability of villages to achieve the SDGs, such concerns should be interpreted as both a reminder and a challenge. This is no longer the time to underestimate the potential of the villages by derisively dismissing them as incapable. This has been learned from experience, considering how many parties were pessimistic when discussing the state budget allocation for villages (the Village Fund), claiming that villages would not be able to manage their own budgets. As it turns out, these concerns were successfully answered by the villages' effective management of the Village Fund. Eko (2014) likens the pessimistic view of the village to a rearview mirror and brakes on a car. Gasoline, gears, and bright lights are needed to keep the vehicle moving forward on a dark road. We have to step on the gas first before we apply the brakes and look in the mirror, so the vehicle continues to move forward smoothly to reach its destination.

It takes joint commitment and courage, from village heads, village officials, upper-level governments, as well as village stakeholders, to make the SDGs Desa a common development goal. In addition to the need to mainstream the SDGs Desa in all village activities, the involvement of stakeholders is necessary in the framework of monitoring, evaluating, and perfecting the actions of the SDGs Desa.

Guidance and supervision are important steps in guarding the implementation of the SDGs Desa. Coaching in this case can include capacity building activities, training, technical assistance, and mentoring. Facilitation is not enough if only carried out by the government, but also needs to be supported by other village activists, such as village leaders, NGOs, universities, international institutions, and even the business world. Assistance is important to be carried out in a sustainable and emancipatory manner by local actors, as a regeneration effort to maintain the pace of village development. The power of the initiatives and innovations of local cadres is a cultural capital for the village in the future. This means village development is more likely to reap success by focusing on the strength of local initiatives, because it is based on experience and knowledge of the needs of local communities. On the other hand, the bureaucratic elite's ideas, which are packaged in government programs, are not necessarily able to produce meaningful changes.

Achieving the SDGs Desa by 2030 in the 75,265 villages throughout Indonesia is not an easy task, but it is not impossible to achieve. For this reason, all resources must be mobilized together; synergy is essential, as is mutual cooperation, working hand-in-hand, throughout the entire process and implementation of development.

The strategy of localizing the SDGs to villages in Indonesia since 2021 has reaped its benefits in 2024. At the village level, there are 12 SDGs Desa goals achieved in 1,259 villages. This results in optimism that the 2030, SDGs Desa target can be achieved. The objectives of the SDGs Desa that have been achieved are:

1. Goal 1: Village without Poverty: 3 Villages.
2. Goal 3: Healthy and Prosperous Village: 3 Villages.
3. Goal 5: Involvement of Village Women: 3 Villages.
4. Goal 6: Villages with Clean Water and Sanitation: 1 Village.
5. Goal 7: Clean and Renewable Energy Villages: 1,109 Villages.
6. Goal 9: Needs-Based Village Infrastructure and Innovation: 79 Villages.
7. Goal 11: Safe and Comfortable Village Residential Area: 1 Village.
8. Goal 12: Environmentally Conscious Village Consumption and Production: 2 Villages.
9. Goal 13: Climate Change Responsive Villages: 41 Villages.

From the many best practices and lessons learned about village development, the main determinant of success is collaboration. The progress and achievements of the village cannot be separated from the support of cooperation with various parties. This means the achievement of the SDGs Desa will also depend on strong commitment and synergy from all parties.

The issue of collaboration is of vital importance because, whether recognized or not, villages have limited resources in terms of both human capacity, such as experts, and non-human capacity, such as finance. For example, universities can provide various consultations and training courses to improve the competence of village officials, including public service management. Universities can also improve the community's capacity, for example, in terms of managing economic enterprises. The private sector, often in the form of corporate social responsibility, also supports village development. Moreover, collaboration is needed when the village is required to make adjustments to reforms on public service management, for example, when budgeting requires operationalization techniques for accountability and transparency. Likewise, collaboration with NGOs can provide training and assistance to the community and village officials. Assistance from NGOs is a good choice for villages when they expect flexibility, not bureaucratic rigidity, making it easier for village officials and villagers to get help from NGO experts and volunteers.

Awareness and involvement of all relevant parties to work hand-in-hand and collaborate is a strategic key to achieving the SDGs Desa by 2030. No one should be left behind and sit idly by, let alone think the burden of achieving the SDGs Desa lies only on the shoulders of the government. All components of the nation must synergize, work together, and campaign, so that all people are called on to participate in achieving the SDGs Desa. This is because the success of Indonesia's development is an accumulation of the successful development of the villages throughout the nation. This view is justified, because most of the nation's weaknesses, as well as its sources of wealth, are in the village.

As part of a global development action, Indonesia has an interest in accelerating the achievement of the SDGs, and that must start from the village. Small steps, which are focused and measurable, and gain the support of multiple parties can then move together, having a giant impact on development.

Let's work from the villages for Indonesia. Let's build Indonesia from the villages. Advance the village, advance Indonesia.

References

Eko, S. 2014. *Desa Membangun Indonesia* (Villages Building Indonesia). Yogyakarta: FPPD.

Statistics Indonesia (BPS). 2020a. *Statistik Kriminal Tahun 2019*. BPS. Jakarta: BPS.

Statitics Indonesia (BPS). 2020b. *Tingkat Ketimpangan Pengeluaran Penduduk Indonesia Maret 2024*. Jakarta: BPS

United Nations. 2015. *Transforming Our World: The 2030 Agenda For Sustainable Development*. New York: UN.

Yustika, A.E. 2016. *Pengembangan Sumber Daya Perdesaan yang Berkelanjutan dan Berkearifan Lokal*. Speech to the Open Senate Meeting to Commemorate the 53rd Anniversary of Jenderal Soedirman University, 23 September 2016.

Bibliography

Abdulai, D., O. Knauf and L. O'Riordan. 2020. *Achieving Sustainable Development Goals 2030 in Africa: A Critical Review of the Sustainability of Western Approaches* In S.O. Idowu, R. Schmidpeter, L. Zu, (eds.), The Future of the UN Sustainable Development Goals: Business Perspectives for Global Development in 2030. Switzerland: Springer.

Agusta, I. 2014. *Diskursus, Kekuasaan, dan Praktik Kemiskinan di Pedesaan* (Discourse, Power, and the Practice of Poverty in Rural Areas). Jakarta: Yayasan Obor Indonesia.

Agusta, I. and A.M. Soleh. 2004. *Jejak-jejak Kesejahteraan: Evaluasi Benefit Program Pembangunan Prasarana Pendukung Desa Tertinggal* (Footprints of Welfare: Evaluation of the Benefits of the Supporting Infrastructure Development Program for Underdeveloped Villages). Jakarta: Binasiamindo Aneka.

Alatas, S.F. 2010. *Diskursus Alternatif dalam Ilmu Sosial Asia: Tanggapan terhadap Eurosentrisme.* Indonesian translation from Alternative Discourses in Asian Social Science: A Response to Eurocentrism. Bandung: Mizan Media Utama.

Alisjahbana, A.S. and E. Murniningtyas. 2018. *Tujuan Pembangunan Berkelanjutan di Indonesia: Konsep Target dan Strategi Implementasi* (Sustainable Development Goals in Indonesia: Target Concept and Implementation Strategy). Bandung: Unpad Press.

Amalik, M.J. 2013. *Pemberdayaan Masyarakat Desa dalam Ayunan Politik* (Empowering Village Communities in Political Swings). Jurnal Mandatory, 10 (1).

Arnstein, S.R. 1969. *A Ladder of Citizen Participation.* Journal of the American Institute of Planners, 35 (4).

Basile, E. and I. Baud. 2019. *Introduction: Reflecting on Development Studies in the New Millennium.* In I. Baud, E. Basile, T. Kontinen and S.V. Itter, (eds.). Building Development Studies for the New Millennium. Cham, Switzerland: Palgrave Macmillan.

Bellah, R.N. 1992. *Religi Tokugawa: Akar-akar Budaya Jepang.* Indonesian translation from Tokugawa Religion: The Values of Pre-Industrial Japan. Jakarta: Gramedia.

Berger, P.L. 2005. Piramida Kurban Manusia: Etika Politik dan Perubahan Sosial. Indonesian translation from Pyramids of Sacrifice. Jakarta: LP3ES.

Biermann, F. and N. Kanie. 2017. *Conclusion: Key Challenges for Global Governance through Goals.* In N. Kanie and F. Biermann (eds.). Governing through Goals: Sustainable Development Goals as Governance Innovation. Massachusetts: MIT Press.

Blakely, E.J. and T.K. Bradshaw. 2002. *Planning Local Economic Development: Theory and Practice.* Thousand Oaks, California: SAGE.

Breman, J. and G. Wiradi. 2004. *Masa Cerah dan Masa Suram di Pedesaan Jawa, Studi Kasus Dinamika Sosio-Ekonomi di Dua Desa Menjelang Akhir Abad Ke-20.* Indonesian translation from Good Times and Bad Times in Rural Java. Jakarta: LP3ES.

A. H. Iskandar, *Accelerating the Achievement of the National Sustainable Development Goals in Indonesia*, Engaging Indonesia, https://doi.org/10.1007/978-981-97-8919-1

Bulletin of the Directorate General of Forest Product Management. 2023. *Perdagangan Karbon, Trik Cerdik Kendalikan Iklim* (Carbon Trading: A Smart Trick to Control the Climate). Jakarta: Ministry of Environment and Forestry.

Cernea, M. (ed.). 1985. *Putting People First: Sociological Variables in Rural Development*. New York: Oxford University Press.

Chambers, R. 1987. *Pembangunan Desa: Mulai dari Belakang*. Indonesian translation from Rural Development: Putting the Last First. Jakarta: LP3ES.

Chambers, R. 1994. *PRA (Participatory Rural Appraisal), Memahami Desa Secara Patisipatif*. Indonesian translation from Participatory rural appraisal (PRA): Challenges, Potentials and Paradigm. Yogyakarta: Kanisius.

Coleman, J.S. 1994. *Foundations of Social Theory*. Cambridge: Harvard University Press.

Concern Worldwide and Welthungerhilfe. 2019. *Global Hunger Index: The Challenge Of Hunger And Climate Change*. Dublin: Helvetas.

Crowther, D., S. Seifi and A. Moyeen. 2018. *Responsibility and Governance in Achieving Sustainability*. In D. Crowther, S. Seifi and A. Moyeen, (eds.). The Goals of Sustainable Development: Responsibility and Governance. Switzerland: Springer.

Dahlan, M.A. 1978. *Sosialisasi Pola Hidup Sederhana* (Socialization of Simple Lifestyle Patterns). Prisma, 7 (10).

Duc, A.L. 2017. *Religious Environmentalism and Environmental Sustainability in Asia*. In J. Servaes, (ed.). Sustainable Development Goals in the Asian Context. Singapore: Springer.

Eko, S. and A. Rozaki (eds.) 2005. *Prakarsa Desentralisasi dan Otonomi Desa* (Decentralization and Village Autonomy Initiative). Yogyakarta: IRE Press.

Eko, S. 2014. *Desa Membangun Indonesia* (Villages Building Indonesia). Yogyakarta: FPPD.

Esteva, G. 1992. *Development*. In W. Sachs, (ed.). The Development Dictionary: A Guide to Knowledge as Power. London: Zed Books.

Fan, S. and P. Polman. 2014. *An Ambitious Development Goal: Ending Hunger and Undernutrition by 2025*. In Global Food Policy Report. Washington, D.C.: IFPRI.

Fakih, Mansour. 2002. *Runtuhnya Teori Pembangunan dan Globalisasi* (The Collapse of Development and Globalization Theories). Yogyakarta: Insist Press and Pustaka Pelajar.

Food and Agriculture Organization (FAO). 2008. *An Introduction to the Basic Concepts of Food Security*. Rome: FAO.

Friedmann, J. 1992. *Empowerment: The Politics of Alternative Development*. Cambridge: Blackwell.

Geertz, C. 1986. *Mojokuto: Dinamika Sosial Sebuah Kota di Jawa*. Indonesian translation from The Social History of an Indonesian Town. Jakarta: Grafitipers.

Harcourt, W. 2004. *The Road to the UN Millennium Development Goals: Some Insights into the International Debate*. Accessed from: www.ncdo.nl.

Hermawan, R. 2015. *Desa dalam Kerangka Negara Kesatuan Republik Indonesia: Melihat Desa dari Sudut Pandang Aturan Perundang-Undangan*. In Jurnal Desentralisasi, 13 (1).

Heryanto, A. 1990. *The Making of Language: Developmentalism in Indonesia*. In Prisma: The Indonesian Indicator, 50.

Higgins, B. and D.J. Savoie. 2005. *Regional Development Theories and Their Application*. New Brunswick: Transaction.

Hikmat, R.H. 2001. *Strategi Pemberdayaan Masyarakat* (Community Empowerment Strategy). Bandung: Humaniora.

Husaini, E.W. Widianto, and E.D. Kristanto. 2011. *Aksi Melawan Imperium Modal: Gerakan Rakyat Melawan Pabrik Semen* (Action Against the Capital Empire: People's Movement Against the Cement Factory). Yogyakarta: Resist Book.

Husken, F. 1998. *Masyarakat Desa dalam Perubahan Zaman, Sejarah Diferensiasi Sosial di Jawa 1830–1980*. Indonesian translation from Village Society in Changing Times: A History of Social Differentiation in Java 1830–1980. Jakarta: Gramedia Widiasarana Indonesia.

Indonesian Ministry of Home Affairs. 2007. *Academic Draft of Village Law*. Jakarta: Home Affairs Ministry.

Indonesian Ministry of Villages, Development of Disadvantaged Regions, and Transmigration. 2015. *Developing Villages Index*. Jakarta: Ministry of Villages, Development of Disadvantaged Regions, and Transmigration.

Indonesian Ministry of Marine Affairs and Fisheries. 2016. Kelautan dan Perikanan dalam Angka Tahun 2016 (Marine and Fisheries in Figures 2016. Jakarta: Indonesian Ministry of Marine Affairs and Fisheries.

Indonesian Ministry of Health. 2018. *Potret Kesehatan Indonesia dari Riskesdas 2018* (A Portrait of Indonesian Health from the 2018 Basic Health Research). Jakarta: Ministry of Health.

Indonesia, Republik. 2018. *Indonesia Second Biennial Update Report Under the United Nations Framework Convention on Climate Change*. Jakarta: Republik Indonesia.

Institute for Essential Services Reform. 2019. Indonesia Clean Energy Outlook. Jakarta: IESR.

Israel, A. 1992. *Pengembangan Kelembagaan Pengalaman Proyek-Proyek Bank Dunia*. Indonesian translation from Institutional Development: Incentives to Performance. Jakarta: LP3ES.

Iskandar, A.H. 2020a. *Desa Beranda Depan Pembangunan* (Village, Home Front of Development). www.Jawa.Pos.com. Accessed from: www.jawapos.com/opini/11/07/2020/desa-beranda-depan-pembangunan.

Iskandar, A.H. 2020b. *Kebijakan Strategis Pemberdayaan Masyarakat Menuju Kemandirian Desa* (Strategic Policy for Community Empowerment Towards Village Autonomy). Scientific speech to obtain an Honorary Doctorate Degree in Community Empowerment Management, delivered at Open Senate Meeting of Yogyakarta State University.

Kanie, N., S. Bernstein, F. Biermann and P.M. Haas. 2017. *Introduction: Global Governance through Goal Setting*, in N. Kanie and F. Biermann, (eds.). Governing through Goals: Sustainable Development Goals as Governance Innovation. Massachusetts: MIT Press.

Kartohadikoesoemo, S. 1984. *Desa* (Village). Jakarta: Balai Pustaka.

Koentjaraningrat, R.M. 1971. *Manusia dan Kebudayaan di Indonesia* (Humans and Culture in Indonesia). Jakarta: Djambatan.

Korten, D.C. 2001. *Menuju Abad Ke-21: Tindakan Sukarela dan Agenda Global* (Coming to Terms with the 21st Century: Voluntary Action and the Global Agenda). Indonesian translation. Jakarta: Yayasan Obor Indonesia.

Kroupova, Z., G. Cervena and M. Antouskova. 2009. *Tools for Agriculture and Rural Development Policy Evaluation*. In The Business Review, 14 (1).

Laksono, P.M. 2009. *Tradisi dalam Struktur Masyarakat Jawa Kerajaan dan Pedesaan, Alih Ubah Model Berpikir Jawa* (Tradition in the Structure of Javanese Royal and Rural Societies, the Transformation of Javanese Thinking Models). Yogyakarta: Kepel Press.

Li, T.M. 2002. *Keterpinggiran, Kekuasaan dan Produksi: Analisis terhadap Transformasi Daerah Pedalaman* (Marginality, Power and Production: Analysing Upland Transformations). In T.M. Li, (ed.). Proses Transformasi Daerah Pedalaman di Indonesia (Transforming the Indonesian Uplands: Marginality, Power and Production). Jakarta: Yayasan Obor Indonesia.

Mariana, D. 2013. *Dana Alokasi Desa: Terobosan Kebijakan Distribusi Keuangan Negara untuk Desa* (Village Allocation Fund: A Breakthrough in State Financial Distribution Policy for Villages). In Jurnal Mandatory, 10 (1).

Maryaningsih, N., O. Hermansyah and M. Savitri. 2014. *Pengaruh Infrastruktur terhadap Pertumbuhan Ekonomi Indonesia* (The Impact of Infrastructure on Indonesia's Economic Growth). In Buletin Ekonomi Moneter dan Perbankan, 17 (1).

McMichael, P. 2003. *Development and Social Change: A Global Perspective*. Thousand Oaks: Pine Forge Press.

MindFuture Foundation. 2020. *Life with Artificials: the 18th Sustainable Development Goal*. Denmark: MindFuture Foundation.

Mikkelsen, B. 1999. *Metode Penelitian Partisipatoris dan Upaya-Upaya Pemberdayaan*. Indonesian translation from Methods for Development Work and Research: A Guide for Practitioners. Jakarta: Yayasan Obor Indonesia.

Moore, B. J. 1966. *Social Origins of Dictatorship and Democracy: Lord and Peasant in the Making of The Modern World*. Boston: Bacon Press.

Mubyarto. 1987. *Ekonomi Pancasila: Gagasan dan Kemungkinan* (Pancasila Economics: Ideas and Possibilities). Jakarta: LP3ES.

Mubyarto. 1996. *Menuju Gerakan Nasional Penanggulangan Kemiskinan* (Towards a National Poverty Alleviation Movement). Yogyakarta: Aditya Media.

Mursanti, E. and F. Tumiwa. 2019. *Strategi Penyediaan Akses Listrik di Perdesaan dan Daerah Terpencil di Indonesia* (Electricity Access Provision Strategy in Rural and Remote Areas in Indonesia). Jakarta: Institute for Essential Services Reform.

National Development Planning Agency (Bappenas) and Statistics Indonesia (BPS). 2015. *Indeks Pembangunan Desa (IPD) 2014: Tantangan Pemenuhan Standar Pelayanan Minimum Desa* (Village Development Index 2014: Challenges in Meeting Minimum Village Service Standards). Jakarta: Bappenas and BPS.

National Development Planning Agency (Bappenas). 2019. Tujuan 14: *Ekosistem Lautan* (Goal 14: Life Below Water). Jakarta: Bappenas.

National Energy Council. 2017. Indonesia National Energy Outlook 2017. Jakarta: Secretariate of National Energy Council.

National Energy Council. 2023. Annual Report 2023. Jakarta: Secretariate of National Energy Council.

Oxfam. 2020. *World's Billionaires Have More Wealth than 4.6 Billion People.* Accessed from: www.oxfamamerica.org/press/worlds billionaires-have-more-wealth-than-46-billion-people.

Prabowo, T.A. 2013. *Perencanaan Desa Wisata Berbasis Pembangunan Berkelanjutan* (Sustainable Development-Based Tourism Village Planning). Yogyakarta: Dinadra Creative.

Ragkos, A., G. Koutouzidou, S. Koutsou and D. Roustemis. 2019. *A New Development Paradigm for Local Animal Breeds and the Role of Information and Communication Technologies.* In A. Theodoridis, A. Ragkos and M. Salampasis (eds.). Innovative Approaches and Applications for Sustainable Rural Development: 8th International Conference, HAICTA 2017, Chania, Crete, Greece, September 21–24, 2017, Selected Papers. Switzerland: Springer.

Rubianto, V. 1996. *Paradigma Asia: Pertautan Kemiskinan dan Kereligiusan dalam Teologi Aloysius Pieris* (The Asian Paradigm: The Link between Poverty and Religiousness in the Theology of Aloysius Pieris). Yogyakarta: Kanisius.

Sajogyo. 1977. *Golongan Miskin dan Partisipasinya dalam Pembangunan Desa* (The Poor and Their Participation in Village Development). In Prisma, 6 (3).

Sajogyo. 1997. *Menciptakan Visi Mendukung Pengembangan Kelompok Swadaya Mandiri dalam Gerakan Nasional PPK* (Creating a Vision to Support the Development of Independent Self-Help Groups in the National Subdistrict Development Program Movement). Jakarta: Puspa Swara.

Sajogyo. 2006. *Ekososiologi: Deideologisasi Teori, Restrukturisasi Aksi (Petani dan Perdesaan sebagai Kasus Uji)* (Ecosociology: Deideologization of Theory, Restructuring of Action (Farmers and Rural Areas as Test Cases)). Yogyakarta: Cindelaras Pustaka Rakyat Cerdas.

Sakamoto, H. 2007. *The Dynamics of Inter-Provincial Income Distribution in Indonesia.* Working Paper Series Vol. 2007-25. Kitakyushu: The International Centre for the Study of East Asian Development.

Sanderson, S.K. 2010. *Sosiologi Makro.* Indonesian translation from Macrosociology An Introduction to Human Societies. Jakarta: Rajawali Pres.

Santosa, I.B. 2011. *Laku Prihatin: Investasi menuju Sukses ala Manusia Jawa* (Practicing Mindfulness: Investing for Success in the Javanese Style). Yogyakarta: Memayu.

Schwab, K. 2019. *Revolusi Industri Keempat.* Indonesian translation from The Fourth Industrial Revolution. Jakarta: Gramedia.

Servaes, J. 2017. *Conclusion: Are the SDGs "Sustainable"?.* In Sustainable Development Goals in the Asian Context, edited by J. Servaes. Singapore: Springer.

Shankar, R. and A. Syah. 2001. *Bridging the Economic Divide within Nations: A Scorecard on the Performance of Regional Development Policies in Reducing Regional Income Disparities.* World Bank Policy Research Working Paper No. 2717.

Siregar, B.B. 2001. *Menelusuri Jejak Ketertinggalan, Merajut Kerukunan Melintasi Krisis* (Tracing the Trail of Disadvantage, Weaving Harmony Across Crisis). Bogor: Pusat P3R-YAE.

Soekanto, S. 2001. *Sosiologi Suatu Pengantar* (Sociology An Introduction), fourth edition. Jakarta: Raja Grafindo Persada.

Stange, P. 2009. *Politik Perhatian, Rasa dalam Kebudayaan Jawa* (Politics of Attention, Feeling in Javanese Culture). Yogyakarta: LKiS.

State Electricity Company (PLN). 2017. *Rencana Umum Penyediaan Tenaga Listrik 2017–2026* (General Plan for Electricity Supply 2017–2026). Jakarta: PLN.

Statistics Indonesia (BPS). 2018a. *Survei Sosial Ekonomi Nasional (Susenas) Tahun 2018* (2018 National Socio-Economic Survey). Jakarta: BPS.

Statistics Indonesia (BPS). 2018b. *Susenas MSBP Tahun 2018* (2018 Susenas Social, Cultural and Educational Module). BPS. Jakarta: BPS.

Statistics Indonesia (BPS). 2019a. *Indeks Pembangunan Desa (IPD) 2018* (Village Development Index 2018). Jakarta. BPS.

Statistics Indonesia (BPS). 2019b. *Statistik Kriminal Tahun 2019* (Crime Statistics 2019). BPS. Jakarta: BPS.

Statistics Indonesia (BPS). 2019c. *Survei Sosial Ekonomi Nasional (Susenas) Tahun 2019* (2019 National Socio-Economic Survey). Jakarta: BPS.

Statistics Indonesia (BPS). 2020a. *Laporan Bulanan Data Sosial Ekonomi, Edisi 124, September 2020* (Monthly Socio-Economic Data Report, Issue 124, September 2020). Jakarta: BPS.

Statistics Indonesia (BPS). 2020b. *Luas Lahan Kritis menurut Provinsi dan tingkat Kekritisan* (Critical Land Area by Province and Criticality Level). Accessed from: www.bps.go.id/ indicator/ 60/588/1/luas-lahan-kritis-menurut-provinsi-dan tingkat-kekritisan-lahan.html.

Statistics Indonesia (BPS). 2020c. *Statistik Indonesia Tahun 2020* (Indonesian Statistics 2020). Jakarta: BPS.

Statistics Indonesia (BPS). 2020d. *Tingkat Ketimpangan Pengeluaran Penduduk Indonesia 2020* (Level of Inequality in Indonesian Population Expenditure in 2020). Jakarta: BPS

Supriatna, T. 1997. *Birokrasi, Pemberdayaan dan Pengentasan Kemiskinan* (Bureaucracy, Empowerment and Poverty Alleviation). Bandung: Humaniora Utama Press.

Suwarsono, A, and Y. So. 2014. *Perubahan Sosial dan Pembangunan.* Indonesian translation from Social Change and Development Jakarta: LP3ES.

Syamsu, S. 2008. *Memahami Perkembangan Desa di Indonesia* (Understanding Village Development in Indonesia). In Government: Jurnal Ilmu Pemerintahan. 1 (1).

Tanjung, M.A. et al. 2018. *Arus Baru Ekonomi Indonesia* (New Currents in the Indonesian Economy). Jakarta: Flasmediatama.

Thomas, A. 2002. *Meanings and Views of Development.* In T. Allen and A. Thomas, (eds.). Poverty and Development: Into the 21st Century. Oxford: Oxford University Press.

Togo, K. 2001. *A Brief Survey on Regional Convergence in East Asian Economies.* Musashi University Working Paper 5.

UNDP. 1990. *Human Development Report 1990.* New York: UNDP.

UNDP. 2008. *Target and Indicators.* New York: UNDP.

UNDP. 2019. *Goal 13: Climate Actions.* New York: UNDP

United Nations. 2000. *Progress towards the International Development Goals: A Better World for All.* Washington: UN.

United Nations. 2007. *The Wye Group Handbook: Rural Households' Livelihood and Well-Being: Statistics on Rural Development and Agriculture Household Income.* Geneva: UN.

United Nations. 2015. *Transforming Our World: The 2030 Agenda For Sustainable Development.* New York: UN.

United Nations. 2016. *Roadmap for Localizing the SDGs: Implementation and Monitoring at Subnational Level.* New York: UN.

United Nations. 2019. *The Sustainable Development Goals Report 2019.* New York: UN.

Uphoff, N. 1985. *Fitting Projects to People.* In M. Cernea, (ed.) 1985. Putting People First: Sociological Variables in Rural Development. New York: Oxford University Press.

Verhagen, Koenraad. 1996. *Pengembangan Keswadayaan: Pengalaman LSM di Tiga Negara.* Indonesian translation from Self-help Promotion: A Challenge to the NGO Community. Jakarta: Penebar Swadaya.

Visseren-Hamakers, I.J. 2020. *The 18th Sustainable Development Goal.* In Earth System Governance, 3.

Weber, M. 2004. *Etika Protestan dan Spirit Kapitalisme.* Indonesian translation from The Protestant Ethic and the Spirit of Capitalism. Yogyakarta: Pustaka Pelajar.

Widjojo, Prasetijono et al. 2004. Indonesia. *Laporan Perkembangan Pencapaian Tujuan Pembangunan Milenium* (Progress Report on Achieving the Millennium Development Goals). Jakarta. Bappenas.

World Bank. 1994. *World Development Report: Infrastructure for Development.* New York: Oxford University Press.

Yustika, A.E. 2016. *Pengembangan Sumber Daya Perdesaan yang Berkelanjutan dan Berkearifan Lokal* (Sustainable and Locally Wisdom Based Rural Resource Development). Speech to the Open Senate Meeting to Commemorate the 53rd Anniversary of Jenderal Soedirman University, 23 September 2016.

Yusuf, A.A. 2006. *On the re-assessment of inequality in Indonesia: household survey or national account?* Paper submitted to the 8th IRSA International Conference, 18–19 August, 2006, Malang, Indonesia.

Yusuf, A.A., A. Sumner. 2015. *Growth, Poverty, and Inequality under Jokowi.* In Bulletin of Indonesian Economic Studies, 51.

Index

A. H. Iskandar, *Accelerating the Achievement of the National Sustainable Development Goals in Indonesia*, Engaging Indonesia, https://doi.org/10.1007/978-981-97-8919-1

U

V

W

MIX
Papier aus verantwortungsvollen Quellen
Paper from responsible sources
FSC® C105338
FSC
www.fsc.org

If you have any concerns about our products,
you can contact us on
ProductSafety@springernature.com

In case Publisher is established outside the EU,
the EU authorized representative is:
Springer Nature Customer Service Center GmbH
Europaplatz 3, 69115 Heidelberg, Germany

Printed by Libri Plureos GmbH
in Hamburg, Germany